Springer Proceedings in Advanced Robotics 12

The Springer Proceedings in Advanced Robotics (SPAR) publishes new developments and advances in the fields of robotics research, rapidly and informally but with a high quality.

The intent is to cover all the technical contents, applications, and multidisciplinary aspects of robotics, embedded in the fields of Mechanical Engineering, Computer Science, Electrical Engineering, Mechatronics, Control, and Life Sciences, as well as the methodologies behind them.

The publications within the "Springer Proceedings in Advanced Robotics" are primarily proceedings and post-proceedings of important conferences, symposia and congresses. They cover significant recent developments in the field, both of a foundational and applicable character. Also considered for publication are edited monographs, contributed volumes and lecture notes of exceptionally high quality and interest.

An important characteristic feature of the series is the short publication time and world-wide distribution. This permits a rapid and broad dissemination of research results.

More information about this series at http://www.springer.com/series/15556

spar

Federica Ferraguti · Valeria Villani ·
Lorenzo Sabattini · Marcello Bonfè
Editors

Human-Friendly Robotics 2019

12th International Workshop

Springer

Editors
Federica Ferraguti
Department of Sciences and Methods for Engineering
University of Modena and Reggio Emilia
Reggio Emilia, Italy

Valeria Villani
Department of Sciences and Methods for Engineering
University of Modena and Reggio Emilia
Reggio Emilia, Italy

Lorenzo Sabattini
Department of Sciences and Methods for Engineering
University of Modena and Reggio Emilia
Reggio Emilia, Italy

Marcello Bonfè
Department of Engineering
University of Ferrara
Ferrara, Italy

ISSN 2511-1256 ISSN 2511-1264 (electronic)
Springer Proceedings in Advanced Robotics
ISBN 978-3-030-42025-3 ISBN 978-3-030-42026-0 (eBook)
https://doi.org/10.1007/978-3-030-42026-0

This Springer imprint is published by the registered company Springer Nature Switzerland AG
The registered company address is: Gewerbestrasse 11, 6330 Cham, Switzerland

Foreword

At the dawn of the century's third decade, robotics is reaching an elevated level of maturity and continues to benefit from the advances and innovations in its enabling technologies. These all are contributing to an unprecedented effort to bringing robots to human environment in hospitals and homes, factories and schools, in the field for robots fighting fires, making goods and products, picking fruits and watering the farmland, saving time and lives. Robots today hold the promise for making a considerable impact in a wide range of real-world applications from industrial manufacturing to health care, transportation and exploration of the deep space and sea. Tomorrow, robots will become pervasive and touch upon many aspects of modern life.

The *Springer Tracts in Advanced Robotics (STAR)* was launched in 2002 with the goal of bringing to the research community the latest advances in the robotics field based on their significance and quality. During the latest fifteen years, the STAR series has featured publication of both monographs and edited collections. Among the latter, the proceedings of thematic symposia devoted to excellence in robotics research, such as ISRR, ISER, FSR and WAFR, have been regularly included in STAR.

The expansion of our field as well as the emergence of new research areas has motivated us to enlarge the pool of proceedings in the STAR series in the past few years. This has ultimately led to launching a sister series in parallel to STAR. The *Springer Proceedings in Advanced Robotics (SPAR)* is dedicated to the timely dissemination of the latest research results presented in selected symposia and workshops.

This volume of the SPAR series brings a selection of the papers presented at the twelfth edition of the International Workshop on Human-Friendly Robotics (HFR). This symposium took place in Modena, Italy, from October 24 to 25, 2019. The volume edited by Valeria Villani, Federica Ferraguti, Lorenzo Sabattini and Marcello Bonfè is a collection of 13 contributions spanning a wide range of topics related to human–robot interaction, both physical and cognitive, including theories, methodologies, technologies, empirical and experimental studies.

From its classical venue to its program dense of presentations by young scholars, the twelfth edition of HFR culminates with this valuable reference on the current developments and new directions of human-friendly robotics—a genuine tribute to its contributors and organizers!

December 2019

Bruno Siciliano
Oussama Khatib
SPAR Editors

Contents

Guiding Quadrotor Landing with Pointing Gestures

Boris Gromov(✉), Luca Gambardella, and Alessandro Giusti

Dalle Molle Institute for Artificial Intelligence (IDSIA), USI/SUPSI, Lugano, Switzerland
boris@idsia.ch

Abstract. We present a system which allows an operator to land a quadrotor on a precise spot in its proximity by only using pointing gestures; the system has very limited requirements in terms of robot capabilities, relies on an unobtrusive bracelet-like device worn by the operator, and depends on proven, field-ready technologies. During the interaction, the robot continuously provides feedback by controlling its position in real time: such feedback has a fundamental role in mitigating sensing inaccuracies and improving user experience. We report a user study where our approach compares well with a standard joystick-based controller in terms of intuitiveness (amount of training required), landing spot accuracy, and efficiency.

Keywords: Pointing gestures · Human-robot interaction · Quadrotor landing

Videos, Datasets, and Code
Videos, datasets, and code to reproduce our results are available at: http://people.idsia.ch/~gromov/hri-landing.

1 Introduction

Modern quadrotors can perform fully-autonomous missions in unstructured outdoor environments (e.g. for mapping, surveillance, etc); however, one delicate step in which human guidance, or at least supervision, is still necessary is landing, especially if this has to occur in an unstructured or populated environment. The landing spot should be sufficiently flat, dry, reasonably far from obstacles, and free of features such as long grass that would interfere with the rotors. Not surprisingly, many experienced pilots choose to land small quadrotors directly on their hand—a solution that is unsafe for the inexperienced operator, and unsuitable for larger quadrotors. In general, landing a quadrotor on an unpaved, unstructured surface requires a careful and critical choice of the landing spot by the operator. While approaches to automatically identify such potential landing spots by means of on-board sensing have been proposed in the literature [9,21],

F. Ferraguti et al. (Eds.): HFR 2019, SPAR 12, pp. 1–14, 2020.
https://doi.org/10.1007/978-3-030-42026-0_1

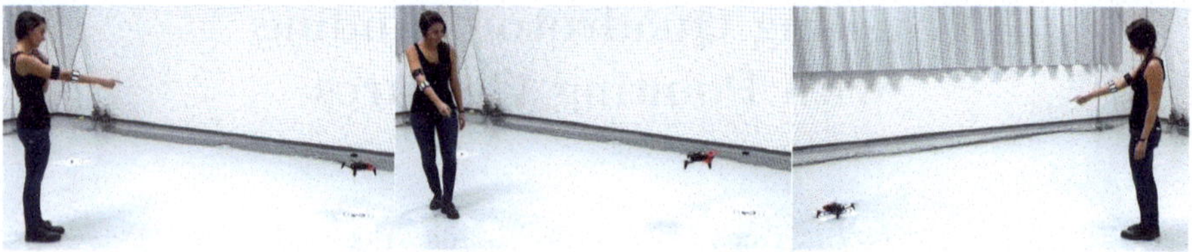

Fig. 1. One of the subjects pointing at the drone to select it (*left*), then guiding it (*center*) to land on a target (*right*).

it is realistic to assume that in many real-world scenarios the guidance of an operator in close proximity of the area will still be required, e.g. for avoiding puddles—which may look like perfectly-flat spots in a 3D reconstruction.

The standard approach for landing a quadrotor is by guiding it using a joystick; this is an efficient technique but requires the operator to be trained, and requires the use of both hands. We target scenarios in which the operator is not necessarily a trained pilot and may be in fact busy with other tasks. Such an example could be found in the future of search and rescue missions with mixed teams of humans and robots: a rescuer might need to land a quadrotor in its vicinity (for changing a battery, retrieving a carried object, etc.) without having a specific training; in this context, we assume that all potential operators wear an unobtrusive, networked bracelet-like device (e.g. a smartwatch) and that they can take control of a nearby drone to guide it to a safe landing spot.

Another realistic application is a drone delivery. Most part of such a mission can be performed autonomously without human intervention: the drone takes off and follows a set of way points to reach a house of the recipient, using for example Global Positioning System (GPS); however, the GPS localization accuracy in urban environments can be deteriorated and will not allow the drone to land autonomously in a safe manner. In this case, the recipient could guide the drone to a safe landing spot with a pointing gesture.

We focus on this well-defined task and aim at providing a control modality with the following characteristics.

- *Practical/pragmatic:* has no strong requirements on the drone or infrastructure capabilities and can be robustly applied to real-world systems in many realistic conditions (outdoors, indoors).
- *Intuitive:* operators need minimal training to use it.
- *Efficient:* an operator can land a drone in a short time.

The interaction begins when the drone approaches a designated position after completing an autonomous part of a mission. To initiate the landing procedure an operator, e.g. a rescuer or a parcel recipient, points at the drone to select it. In turn, the drone provides a visual feedback to confirm that the control has been transferred to the operator, e.g. by performing a predefined motion primitive or by blinking with its on-board lights. From that moment, the drone follows and hovers over the location on the ground pointed by the operator. Once the operator is determined to land the drone, they simply keep the arm still for a predefined time. A countdown timer starts and the system sends periodic feedback to the user to notify them that the drone is about to land. In case the operator decides to adjust the landing spot, they simply point at another location, the countdown timer cancels and the landing procedure starts over again. This interaction sequence is shown in brief in Fig. 1, please refer to the linked video for details.

In order to prove the viability of this control approach, we implemented it and experimentally compared its performance with joystick-based control.

In the following sections we review the related work (Sect. 2), define the major abstract functionalities required to realize the given interface and task, and discuss implementation options (Sect. 3). We describe our implementation of these functionalities in Sect. 4. To compare our approach with the classic joystick-based control we set up an experiment, which is described in detail in Sect. 5, the results are then analyzed in Sect. 6. Finally, in Sect. 7 we draw the conclusions and describe the future work.

2 Related Work

Since pointing gestures are such a compelling solution to many human-computer and human-robot interaction problems, significant research efforts have been devoted to this topic. To the best of our knowledge, however, this work is the first to approach the issue of landing a drone by using pointing gestures.

There are many works that use iconic gestures [22] to control drones [17,20]. These can use hands, arms, or full-body postures to give discrete commands to the drones, such as "go up", "go down", "turn left", "take off", etc. Although these gestures may be represented by a pointing hand or arm, the exact direction of this pointing is not important. On the contrary, we are interested in pointing gestures that indicate precise directions and locations with respect to the user. Therefore, below we only review the research related to this particular type of pointing gestures and omit the interfaces based on iconic gestures.

Using pointing gestures as an input interface dates back to 1980s, when Bolt presented his now famous work "Put-that-there" [2]. A multi-modal human-computer interaction interface developed in that work was used to manipulate virtual objects on a screen. The input interface consisted of a commercial speech recognition system and a pose sensing device. The pointing device included a stationary transmitter of a nutating magnetic field and a small wired sensor placed on the user's wrist. Altogether the system allowed to manipulate objects using simple voice queries like "Put that there", where "that" would be supported by one pointing gesture and "there" by another. In this case, Bolt argues, the user even does not have to know what the object is or how it is called.

In HRI literature, pointing gestures are often used for pick-and-place tasks [4–6], labeling and/or querying information about objects or locations [4], selecting a robot within a group [16,19], and providing navigational goals [1,12,14,25,26].

One important issue to be solved in natural human-robot interaction that involves pointing is a perception of the user's gestures. This can be a responsibility of a robot, i.e. the recipient of the message, as well as of a group of cooperatively-sensing robots [19]; of the environment [27]; or, as in our case, of a device worn by the user [12,23,26]. The first approach is the most popular in HRI research. On one hand, it is natural because it mimics what humans do when communicating with each other (the recipient of the message is the one which perceives it). On the other hand, it presents important challenges to solve the perception problem, and requires the robot to consistently monitor the user. Relying on sensors placed in the environment relaxes the requirements on the robots, but limits the applicability of the system to properly instrumented areas; in both cases, the positions being pointed at need to be inferred by external observation, which is typically performed with cameras or RGB-D sensors.

2.1 Providing Navigational Goals

We now focus our review on pointing gestures for robot guidance.

Van den Bergh [25] used pointed directions to help a ground robot to explore its environment. The robot continuously looks for a human in its vicinity and once detected begins the interaction. Using an RGB-D sensor (Microsoft Kinect) the system detects the human's hand and wrist. A vector connecting the center of the hand and the wrist is then projected on the ground, giving a principal exploration direction. Finally, the next exploration goal is automatically selected from a set of possible goals with respect to an instantaneous occupancy grid acquired by the robot. The authors also suggest an alternative method to estimate pointing directions, namely a line connecting the eyes and the fingertip, however they do not elaborate on this approach.

Similarly to the previous work, Abidi et al. [1] use a Kinect sensor to extract pointed directions. Navigation goals are continuously sent to the ground robot, which reactively plans its motion and thus allows the user to correct her input on the fly. The main drawback, however, is that the robot has to "keep an eye" on the user in order to reach the final target. To estimate pointed locations authors suggest two approaches: (1) a vector originating from the elbow and passing the hand/finger, and (2) a vector originating from the eyes and also passing the hand/finger. The approaches were compared in a user study, but the only reported result is a subjective satisfaction level of the participants. The majority (62%) preferred the second approach.

Jevtic et al. [14] experimentally compared several interaction modalities in the user study of 24 participants. A ground robot equipped with a Kinect and other sensors was used. The study compares three interaction modalities: direct physical interaction (DPI), person following, and pointing control in area- and waypoint-guidance tasks. The DPI modality requires the user to push the robot by hands, the torques generated at motors are measured via electrical current and then are fed to a friction-compensation controller that drives the robot in the appropriate direction. The person following modality makes the robot to follow the user at a safe distance, the user can stop the robot at any time by raising their left hand above the left elbow and thus can control the robot's precise location. The pointing modality allows the user to command the robot's position with a pointing gesture, where the target location is calculated from the intersection of the ground plane with a line passing through the right elbow and the right hand of the user. The authors measured task completion times, accuracy, and workload (with NASA-TLX questionnaire). Reported results show that the DPI modality is systematically better than the other modalities for all the metrics, while the pointing control shows the worst results.

Such a low performance of the pointing interface used in the study by Jevtic et al. [14] can be explained by a lack of appropriate feedback and a time-sparse nature of the implemented gesture control: the user issues a single command to drive the robot to a goal and see where the system "thinks" they were pointing at only when the robot reaches the target, therefore, the user is unable to efficiently correct the robot's position. These problems are further aggravated by an inherently limited precision of a chosen pointing model (elbow-hand). As reported by many other works (see [1,6,18]), including those from the psychology research (see [13,24])—a more appropriate model would be a line that passes through the head and the fingertip. Also note that contrary to the implemented pointing control, the DPI and person following modalities work in the tracking mode, that provides immediate feedback and allows the user to correct robot's position in the real time. As will be seen later, in our work we mitigate aforementioned flaws by making the robot to instantaneously follow to pointed locations, and thus providing a real time feedback to the user. In our recent work [10] we systematically compared users performance in a pointing task with and without the visual feedback: we have shown that the lack of visual feedback results in significant errors.

2.2 Wearable Sensors

Wearable sensors are an alternative approach to the problem of perceiving pointing gestures. Sugiyama et al. [23] developed a wearable visuo-inertial interface for on-site robot teaching that uses a combination of monocular camera and inertial measurement unit (IMU) to capture hand gestures in the egocentric view of the user. The camera is also used for a monocular simultaneous localization and mapping (SLAM), which allows to localize the user with respect to a common with the robot coordinate frame.

Wolf et al. [26] suggest a gesture-based interface for a robot control, that is based on a device they call BioSleeve—a wearable device placed on the user's forearm and comprised of a set of dry-contact surface electro-myography sensors (EMGs) and an IMU. Optionally, the authors suggest to strap an additional IMU sensor on the upper arm to be able to perform a model-based arm pose reconstruction for pointing gestures. However, no information is given on how a user would localize themselves with respect to the robot in order to control its position.

2.3 Pointing Direction Estimation

Regardless on the specific approach adopted for sensing, assuming a perfect knowledge of the user's posture, one has to solve the problem of interpreting such posture to map it to the point in the environment that the user wants to indicate. This problem has been extensively studied in the psychology research [13,24] which suggests two main models: (a) the *forearm* model assumes that the point lies along the 3D line defined by the axis of the forearm of the pointing arm; (b) the *head-finger* model assumes that the point lies along the line connecting the dominant eye and the tip of the finger. The choice of a particular model in robotic applications mainly depends on the technology available for sensing the user's posture and on the task.

3 Model

In order to realize the task we have defined in the introduction, it is necessary to address the following problems: (a) estimation of the pointed directions and locations with respect to the human, (b) identification and localization of a robot being pointed at, and (c) detection of discrete triggering events that would dispatch commands to the robot.

Pointed Direction. The pointed direction is recovered as a ray in 3D space, expressed in a human-centered reference frame. The frame is fixed while the interaction occurs and has its origin at user's feet; its xy-plane is aligned with the world's horizon; the remaining degree of freedom (a rotation around the vertical axis) is a free parameter. Without loss of generality, we can assume the x-axis is the heading of the first pointing gesture, i.e. the one used to select the robot.

A forearm-mounted IMU is a viable option to meet this requirement as long as it can return accurate relative orientation data without significant drift for the duration of the interaction, and can reliably estimate the vertical direction. A more sophisticated approach may use additional sensors, e.g. multiple IMUs [12, 26], to model more accurately the arm kinematics. Since the IMUs provide only a 3D-rotation, one will also need to acquire the position of the user's shoulder with respect to the human frame. It can be measured directly or estimated using a simple calibration procedure.

Robot Identification and Pose Reconstruction. Since pointed direction is expressed in human's reference frame, it is necessary to find a coordinate transformation between the human and a robot. We require that while a robot is being pointed at, the system can detect it, identify it, and recover its 6D-pose (3D-position + 3D-orientation) with respect to the human-centered frame.

In practice, this can be achieved in numerous ways: with a camera pointed in the same direction as the forearm [12], which can detect the robot's presence and identify its pose using pattern recognition techniques, e.g. relying on robot-mounted visual fiducial markers [8], or detecting active LEDs [7].

Recently we described another efficient method to co-localize a user and a robot [11]. The method relies on synchronized motion of user's arm and the robot: the user points and keeps following a moving robot they want to interact with for a few seconds; the system collects synchronized pairs of pointing rays expressed in user's frame and robot's positions expressed in its odometry frame, the algorithm then finds the coordinate transformation between the human and the robot that fits the captured movements the best.

Triggering. We assume that the system implements a mechanism to trigger the first and second pointing events (at robot and at target) and thus advance the interaction.

A trivial approach is to use a push button on the wearable/handheld device, which however prevents hands-free operation. Other realistic triggering mechanisms include gestures, automatic detection of the start of a pointing gesture [18], fixed time delays [5], or speech [12].

4 Implementation

We implemented a pointing-based interface that consists of two Myo armbands, respectively placed on the upper arm and the forearm. We use a single sensor for the arm pose estimation, however we use both for the gesture detection which is described later.

Myo is an integrated wireless wearable sensor from Thalmic Labs[1] comprised of 9-DoF IMU, eight surface electromyography sensors (EMG) and a processing unit. The device internally fuses the IMU data into an accurate absolute 3D-orientation (roll, pitch, and yaw angles) which we use for the arm pose estimation. The inertial data is transmitted to the host PC over the Bluetooth LE link at 50 Hz rate.

Pointed Direction. To estimate pointed directions we employ a *head-finger* model—a popular choice in robotics [5,6,15,18]—which requires the knowledge of the arm and head positions. However, we simplify this model and assume that: (a) the head position is fixed with respect to the body; (b) the user always points with the straight arm, and (c) the shoulder length (distance from the neck to the shoulder joint) is zero.

Although these simplifications lead to higher lateral and radial static errors, a live update of the drone's position allows to efficiently mitigate them.

In this work we consider the arm as a single link with a 3-DoF ball joint (shoulder) connected to a fixed vertical link (torso). We acquire the orientation of the shoulder with a help of a single Myo armband placed on the forearm.

[1] Myo has been discontinued as of Oct 12, 2018.

Once we know the pointing direction with respect to the human's shoulder we can simply find the pointed location as an intersection of a line and the ground plane.

Robot Identification and Pose Reconstruction. Since there is only one robot to control, there is no need for robot identification.

In order to determine the robot's location with respect to the operator, we assume that the robot is flying at a known altitude over a flat floor; under this assumption, the distance to the robot can be estimated from the pointing ray alone, by intersecting it with the horizontal plane on which the robot is flying.

The only remaining parameter is the relative heading between the operator and the drone. Most of the drones are equipped with an IMU for stabilization purposes, and usually include a magnetometer—the sensor that estimates the heading with respect to the Magnetic North, an absolute reference frame. Therefore, the rotation between the frames is defined by the difference between the human's and robot's absolute headings.

Triggering. Using data from two wireless IMU sensors worn on the arm and forearm, we follow a detection-by-classification paradigm and use a 1D convolutional neural network as a binary classifier. Given the data acquired in the last few seconds, the network predicts whether a pointing gesture occurred in this interval. The network has been trained using data acquired from multiple users, who were prompted by the system to perform the gesture at specific times [3].

Therefore, the action is triggered immediately once the user performs the pointing gesture.

5 Experimental Setup

To confirm the viability of the proposed interface in a guided landing task, we set up an experiment where the users are required to land a quadrotor (Parrot Bebop 2) at a given location using two different interfaces: pointing gestures and a regular joystick.

The experimental environment represents a flying arena with four predefined targets. The targets are placed at the corners of a square with an edge of 3.6 m and numbered in clockwise order. The sequence of the targets is predefined as 1–2–4–3–1, i.e. edge segments alternate with diagonal ones. The subjects were asked to stay in the middle of arena, however they were allowed to step aside to avoid collisions with the drone.

The arena is equipped with the Optitrack motion capture system that provides precise information of the drone's position. This information is used, both, to control the safety margins and to implement autonomous flights. Note, that in general the robot is localized in an arbitrary frame, e.g. in its odometry frame or, as in our case, in the motion capture frame; however, the location of the operator with respect to robot's frame is not known until the "Robot identification and pose reconstruction" step has taken place.

The drone closed-loop controller is built around `bebop_autonomy`[2] ROS-package and accepts velocity and 6D-pose commands. While the joystick interface generates velocity commands, the pointing interface supplies the pose commands.

5.1 Subjects

Five people between 25 and 36 years old have volunteered to participate in the experiment. Majority have reported either no experience in piloting RC-vehicles or a little experience ("tried it a few times").

We conducted two sessions per person: with the joystick and with the pointing-based interface, each consisting of three runs. Each run starts and ends at target 1 and therefore provides four segments. This totals to 12 segments per person or 120 target-to-target segments for all the participants for both interfaces. Three subjects started with the pointing interface and the rest with the joystick.

Prior to each experimental session, individually for each subject, we conducted a training session with the same interface that they were given later. Each training session consisted of two runs.

Two training sessions plus two experimental sessions took approximately 40 min per person, including all the explanations, service times (e.g. replacing the drone's battery), etc.

5.2 Experimental Sequence

At the beginning of each session the drone is placed at target 1. Once the supervisor starts the session the drone takes off automatically. Once it is airborne and stable, it aligns itself with the first target of the segment and turns its back to the user, such that the user's controls, both, for the joystick and for the gestures, are aligned, meaning, e.g., that pushing joystick forward would drive the drone away from the user, in the direction they look to. This way we ensure that the landing errors are not accumulated over the course of experiment and that all the subjects start in equal conditions, both, when controlling the drone with pointing gestures and with a joystick.

From this moment the drone is ready to interact with the user. Once the user lands the drone, it will perform the automatic take off procedure with a delay of 5 seconds from the moment it has been landed. This procedure repeats until the list of targets is exhausted.

Since the interaction pattern with the joystick and the gesture-based interface are slightly different we report them separately.

Pointing-Based Interface. The user points at the floor beneath the drone, i.e. at the crosshair of the target, to select it[3]. The Myo on the user's arm vibrates

[2] http://wiki.ros.org/bebop_autonomy.

[3] Although the drone can be selected in the air, this brings additional error to the relative localization and may deteriorate user experience.

and the drone 'jumps' to signify that it is now being controlled by the user. At the same time, the control station gives a voice feedback through a loudspeaker, telling the user the next target they should bring the drone to. Immediately after that the drone starts to continuously track a newly given location. Once the user is ready to land the drone, they have to maintain the pointed location for approximately half a second. The system starts to count down and makes the upper arm Myo to vibrate every second. The user has about 3 s to change their mind. To adjust the landing position they just have to start moving the arm away and the countdown will be canceled.

Joystick. The behavior of the system in this mode is similar, however the drone is selected automatically and performs the same 'jump' motion as in pointing-based interaction mode, meanwhile the control station gives a voice feedback with the next target number. The user then moves the drone to the required target and presses the button on the joypad to land the drone.

5.3 Performance Metrics.

We define a set of performance metrics to compare the performance of two interfaces:

- *Landing error.* Euclidean distance between the requested landing target P_i and the actual position p_a the drone have landed to: $\varepsilon = |p_a - P_i|$.
- *Time to target.* Time that passed from the moment t_0 the drone has moved 20 cm away from its starting pose till the moment t_1 the user gave command to land it: $\tau = t_1 - t_0$.
- *Trajectory length.* Line integral of the trajectory curve from the start position P_{i-1} to the actual landing position p_a: $\rho = \sum_{k=1}^{N} \|p_{k+1} - p_k\|$, where $p_k \in \{p_1, ...p_N\}$ is a set of all the acquired positions of the drone between P_{i-1} and p_a.

We collect the data with a standard ROS tool `rosbag` and analyze it offline.

6 Results

Landing Error. Fig. 2 reports the landing error metric. We observe no statistically significant difference between the results with the two interfaces; we separately report the error (left pair) and its decomposition in a radial (central pair) and tangential (right pair) component with respect to the user's position. It's interesting to note how the radial component dominates the error in both interfaces, which is expected since the radial component corresponds to the depth direction from the user's perspective: along this direction, the quadrotor's misalignment with respect to the target is much more difficult to assess visually; the landing error is dominated by a perception (rather than control) issue.

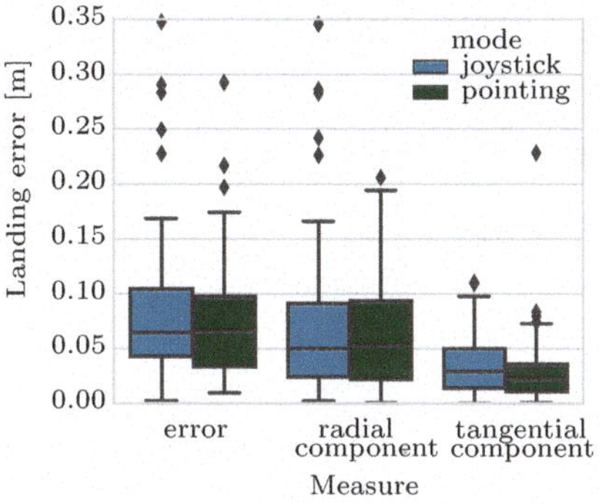

Fig. 2. Statistical analysis of the landing error metric ($N = 60$ for each interface).

Time to Target. Figure 3 (*left*) reports the time to target metric, separately for each of the four segments. We observe that the pointing interface yields a better average performance; the difference is statistically significant under Student's t-test ($p < 0.01$) for 3 of the 4 segments and for the mean over all segments.

Trajectory Length. Figure 3 (*right*) reports the trajectory length metric, separately for each of the four segments. We observe that the pointing interface consistently yields shorter trajectories than the joystick interface; the difference is statistically significant under Student's t-test ($p < 0.01$) for all four segments. On average over all segments, the joystick interface yields a 66% longer trajectory than the straight distance between the targets; the pointing interface yields a 33% longer trajectory.

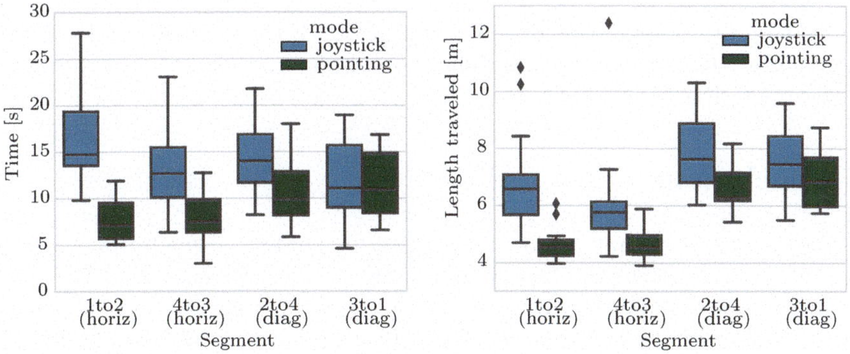

Fig. 3. Statistical analysis of the time to target (*left*) and trajectory length metrics (*right*) ($N = 15$ for each interface and segment).

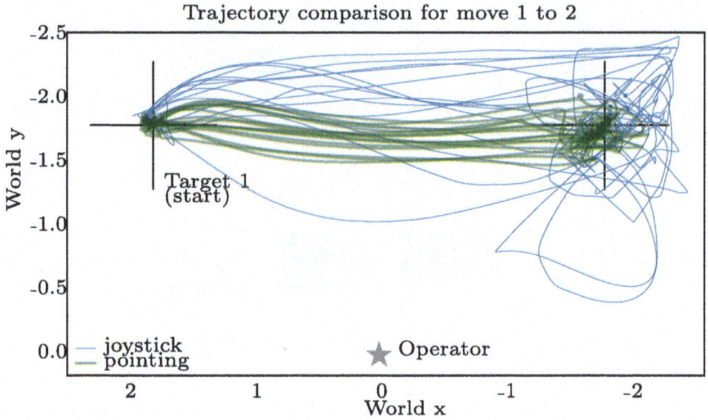

Fig. 4. Comparison of all trajectories flown from target 1 (left) to target 2 (right) for joystick (blue, $N = 15$) and pointing (green, $N = 15$) interfaces.

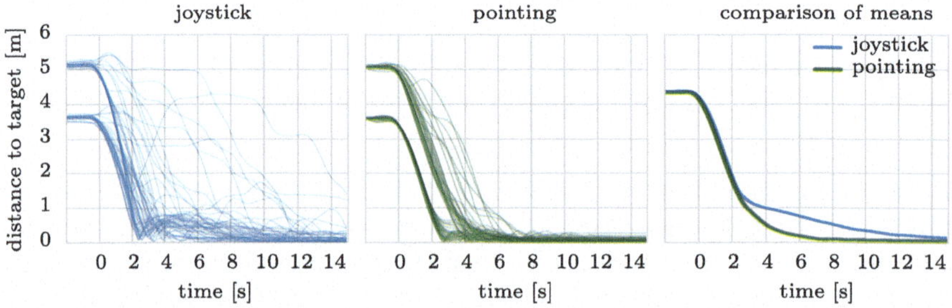

Fig. 5. Evolution in time of the distance to the target, for each trajectory flown: (*left*) joystick interface ($N = 60$), (*center*) pointing interface ($N = 60$), (*right*) average over all trajectories for each interface.

One can also observe this phenomenon in Fig. 4: the trajectories flown with the pointing interface are smoother and more direct and tend to converge faster to the target (Fig. 5).

7 Conclusions

We proposed a novel human-robot interface for landing a quadrotor based on pointing gestures detected by means of unobtrusive wearable sensors. The interface has minimal requirements in terms of robot capabilities, and extensively takes advantage from real-time feedback. In a preliminary user study, it compares favorably with a traditional joystick-based interface in terms of efficiency and intuitiveness.

Acknowledgments. This work was partially supported by the Swiss National Science Foundation (SNSF) through the National Centre of Competence in Research (NCCR) Robotics.

References

1. Abidi, S., Williams, M., Johnston, B.: Human pointing as a robot directive. In: ACM/IEEE International Conference on Human-Robot Interaction, pp. 67–68 (2013)
2. Bolt, R.A.: "Put-that-there": voice and gesture at the graphics interface. In: Proceedings of the 7th Annual Conference on Computer Graphics and Interactive Techniques - SIGGRAPH 1980, pp. 262–270 (1980)
3. Broggini, D., Gromov, B., Gambardella, L.M., Giusti, A.: Learning to detect pointing gestures from wearable IMUs. In: Proceedings of Thirty-Second AAAI Conference on Artificial Intelligence, USA, 2018. AAAI Press (2018)
4. Brooks, A.G., Breazeal, C.: Working with robots and objects: revisiting deictic reference for achieving spatial common ground, pp. 297–304. Gesture (2006)
5. Cosgun, A., Trevor, A.J.B., Christensen, H.I.: Did you mean this object? Detecting ambiguity in pointing gesture targets. In: HRI 2015 Towards a Framework for Joint Action Workshop (2015)
6. Droeschel, D., Stückler, J., Behnke, S.: Learning to interpret pointing gestures with a time-of-flight camera. In: Proceedings of the 6th International Conference on Human-Robot Interaction - HRI 2011, pp. 481–488 (2011)
7. Faessler, M., Mueggler, E., Schwabe, K., Scaramuzza, D.: A monocular pose estimation system based on infrared leds. In: 2014 IEEE International Conference on Robotics and Automation (ICRA), pp. 907–913. IEEE (2014)
8. Fiala, M.: Designing highly reliable fiducial markers. IEEE Trans. Pattern Anal. Mach. Intell. **32**(7), 1317–1324 (2010)
9. Forster, C., Faessler, M., Fontana, F., Werlberger, M., Scaramuzza, D.: Continuous on-board monocular-vision-based elevation mapping applied to autonomous landing of micro aerial vehicles. In: 2015 IEEE International Conference on Robotics and Automation (ICRA), pp. 111–118 (2015)
10. Gromov, B., Abbate, G., Gambardella, L., Giusti, A.: Proximity human-robot interaction using pointing gestures and a wrist-mounted IMU. In: 2019 IEEE International Conference on Robotics and Automation (ICRA), pp. 8084–8091 (2019)
11. Gromov, B., Gambardella, L., Giusti, A.: Robot identification and localization with pointing gestures. In: 2018 IEEE/RSJ International Conference on Intelligent Robots and Systems (IROS), pp. 3921–3928 (2018)
12. Gromov, B., Gambardella, L.M., Di Caro, G.A.: Wearable multi-modal interface for human multi-robot interaction. In: 2016 IEEE International Symposium on Safety, Security, and Rescue Robotics (SSRR), pp. 240–245 (2016)
13. Herbort, O., Kunde, W.: Spatial (mis-) interpretation of pointing gestures to distal spatial referents. J. Exp. Psychol.: Hum. Percept. Perform. **42**(1), 78–89 (2016)
14. Jevtić, A., Doisy, G., Parmet, Y., Edan, Y.: Comparison of interaction modalities for mobile indoor robot guidance: direct physical interaction, person following, and pointing control. IEEE Trans. Hum. Mach. Syst. **45**(6), 653–663 (2015)
15. Mayer, S., Wolf, K., Schneegass, S., Henze, N.: Modeling distant pointing for compensating systematic displacements. In: Proceedings of the ACM CHI 2015 Conference on Human Factors in Computing Systems, vol. 1, pp. 4165–4168 (2015)
16. Nagi, J., Giusti, A., Gambardella, L.M., Di Caro, G.A.: Human-swarm interaction using spatial gestures. In: IEEE International Conference on Intelligent Robots and Systems, pp. 3834–3841 (2014)
17. Ng, W.S., Sharlin, E.: Collocated interaction with flying robots. In: Proceedings - IEEE International Workshop on Robot and Human Interactive Communication, pp. 143–149 (2011)

18. Nickel, K., Stiefelhagen, R.: Visual recognition of pointing gestures for human-robot interaction. Image Vis. Comput. **25**(12), 1875–1884 (2007)
19. Pourmehr, S., Monajjemi, V., Wawerla, J., Vaughan, R., Mori, G.: A robust integrated system for selecting and commanding multiple mobile robots. In: Proceedings - IEEE International Conference on Robotics and Automation, pp. 2874–2879 (2013)
20. Sanna, A., Lamberti, F., Paravati, G., Manuri, F.: A Kinect-based natural interface for quadrotor control. Entertainment Comput. **4**(3), 179–186 (2013)
21. Scherer, S., Chamberlain, L., Singh, S.: First results in autonomous landing and obstacle avoidance by a full-scale helicopter. In: Proceedings - IEEE International Conference on Robotics and Automation, pp. 951–956 (2012)
22. Suarez, J., Murphy, R.R.: Hand gesture recognition with depth images: a review. In: RO-MAN 2012, pp. 411–417. IEEE (2012)
23. Sugiyama, J., Miura, J.: A wearable visuo-inertial interface for humanoid robot control. In: ACM/IEEE International Conference on Human-Robot Interaction, pp. 235–236. IEEE (2013)
24. Taylor, J.L., McCloskey, D.: Pointing. Beh. Brain Res. **29**(1–2), 1–5 (1988)
25. Van den Bergh, M., et al.: Real-time 3D hand gesture interaction with a robot for understanding directions from humans. In: Proceedings - IEEE International Workshop on Robot and Human Interactive Communication, pp. 357–362 (2011)
26. Wolf, M.T., Assad, C., Vernacchia, M.T., Fromm, J., Jethani, H.L.: Gesture-based robot control with variable autonomy from the JPL BioSleeve. In: Proceedings - IEEE International Conference on Robotics and Automation, pp. 1160–1165 (2013)
27. Zivkovic, Z., et al.: Toward low latency gesture control using smart camera network. In: 2008 IEEE Computer Society Conference on Computer Vision and Pattern Recognition Workshops, CVPRW 2008, pp. 1–8. IEEE (2008)

Human-Friendly Multi-Robot Systems: Legibility Analysis

Beatrice Capelli(✉) and Lorenzo Sabattini

Department of Sciences and Methods for Engineering (DISMI),
University of Modena and Reggio Emilia,
Via Amendola 2, 42122 Reggio Emilia, Italy
{beatrice.capelli,lorenzo.sabattini}@unimore.it
http://www.arscontrol.unimore.it

Abstract. This paper investigates the concept of legibility of a multi-robot system. Considering a group of mobile robots moving in the environment according to an artificial potential based control law, we study the effect of the choice of the control parameters on the legibility of the system. With the term legibility we refer to the ability of the multi-robot system to communicate to a user: in particular, we consider a user who shares the environment with the robots, and is requested to understand what is the goal position the multi-robot system is moving to. We analyze the effect of the choice of a few design parameters, named *motion-variables*, performing a set of user studies in a virtual reality setup, with experiments based on the central composite design method.

Keywords: Legibility · Multi-robot systems

1 Introduction

Multi-robot systems are composed of multiple robotic units that, exchanging information among each other, are able to achieve some common objective. They have been deeply investigated in the literature, typically with the aim of defining control strategies to allow them to work in a completely autonomous manner [5]. Human interaction and supervision has been studied as well, in particular with several works considering teleoperation of multi-robot systems [4,8,18–20]. Teleoperation is an effective method when the user acts as a supervisor for the multi-robot system, and works in a physically separated environment.

In this paper, we consider a different situation, in which the user shares the environment with the multi-robot system. Following the taxonomy proposed in [11], we are considering a *proximal interaction* scenario. Besides safety considerations, which are out of the scope of the present paper, we believe that communication is a key issue in this scenario. In particular, communication is fundamental to let the user provide control inputs to the multi-robot system. For this purpose, several methodologies have been developed, that mainly rely on

F. Ferraguti et al. (Eds.): HFR 2019, SPAR 12, pp. 15–28, 2020.
https://doi.org/10.1007/978-3-030-42026-0_2

gesture, voice, or facial expression recognition [1,15,17], or on the use of input devices such as joysticks or wearable systems [10,25,26].

All these methodologies have the objective to make the intention of the user understandable and exploitable by the robots. The user is forced to utilize specific methods to express her/his intention: in this way, the user becomes *robot-friendly.*

In this paper we want to consider the opposite problem, and reversing the approach, making the multi-robot system *human-friendly.* We are, in fact, interested in defining methods to let the multi-robot system provide information to the user, related to the task that is currently being performed. In particular, we are investigating methods to achieve such objective without the use of explicit cues, such as voice synthesis, intentional movements (e.g., of the head for humanoid robots), monitors, lights, or LED [13,14,21–23]. In this manner, ad hoc communication devices are not necessary: the objective is to define how and when humans can be able to *implicitly* understand relevant information about the task the multi-robot system is performing.

This work builds upon our previous results [2,3], where we proposed the definition of *legibility* of a multi-robot system, as a function of some relevant parameters that characterize the motion of the multi-robot system itself. We investigated three *motion-variables* (trajectory, dispersion and stiffness) and we found out that they are relevant for the communication between a multi-robot system and a user. Specifically, trajectory was found relevant to communicate the goal, while the other two variables allow to improve the rapidity of the communication. As both previous works confirmed that the *minimum-jerk* trajectory is the correct choice for this type of communication, in this paper we investigate only dispersion and stiffness. While in our previous works we experimentally found that the influence of these variables is relevant, we only had simple intuitions about the actual relationship between the value of the variables and the legibility of the system. In this paper we propose a more exhaustive experimental setup, with the objective of further investigating the functional relationship between the *motion-variables* and the legibility of the system.

Defining such a functional relationship would allow us to combine legibility with other functional requirements and constraints for the multi-robot system. In particular, legibility can be considered as a (soft) constraint to be satisfied, while achieving some objective (e.g., exploring an area), and, at the same time, satisfying other constraints (e.g., connectivity preservation).

The rest of the paper is organized as follows. The formal definition of the problem addressed in this paper is provided in Sect. 2. The experimental methodology is then detailed in Sect. 3. Subsequently, the results of the experimental campaign are discussed in Sect. 4. Finally, concluding remarks are given in Sect. 5.

2 Problem Definition

In this paper we address the following problem: *investigating the existence of a functional relationship to map the state variables of the multi-robot system into its legibility.*

Generally speaking, legibility is the ability of a system of being understandable by a user. In particular, in our previous works [2,3], we introduced the concept of legibility of a multi-robot system as the ability of the group of robots to communicate some information to the user without the use of explicit communication.

In our previous works, we represented the state of the system Υ as a set of variables that describe its motion pattern. We referred to these variables as *motion-variables*, defined in details as:

- $\mathcal{T}$: trajectory of the center of the group;
- $\mathcal{D}$: dispersion of the group;
- $\mathcal{S}$: stiffness of the interconnection among the robots in the group.

In the aforementioned previous works, by means of extensive user studies, we found that the influence of the chosen *motion-variables* on the legibility of the system is statistically significant.

We considered the multi-robot system to be possibly composed of multiple groups: we refer with Q to the set of groups, and $q \in Q$ to each single group. We also considered that each group has a goal $\mathcal{G}$, that represents the target position to be reached by the group itself. We use the symbol $\mathfrak{G}$ to represent the set of goals for the set of groups.

Statistical significance of the influence of the *motion-variables* on the legibility of the system implies the existence of inference functions, that represent the most likely goal as a function of the state, namely:

$$\mathcal{I}_{GOAL} : (\Upsilon_q) \rightarrow \mathcal{G} \tag{1}$$

$$\mathbb{I}_{GOAL} : (\Upsilon_Q) \rightarrow \mathfrak{G} \tag{2}$$

In this paper, we aim at characterizing this relationship. For the sake of simplicity of notation, we will hereafter focus on the single group case. However, according to the results reported in [2], everything can be extended to the multiple group case.

Hence, the objective is to characterize the following relationship:

$$\texttt{LEGIBILITY}_{GOAL} : (\Upsilon) \rightarrow \mathbb{R}^+ \tag{3}$$

As in [3], we consider the legibility of the system as the combination of the correct interpretation of the communication and the time in which the communication takes place. In fact, we consider both the correctness of the communication and its rapidity. This function would allow us to characterize the level of legibility associated with a given set of *motion-variables*.

3 Methods

In order to characterize relationship between the *motion-variables* and the legibility, introduced in (3), we propose the use of a response surface design method [16]. In particular, we performed an experimental campaign, conducted in a virtual reality set-up [24], having a set of users performing an experimental task similar to those performed in [2,3].

3.1 Response Surface Design

Response surface design allows to statistically retrieve a model of a response variable with respect to one or more independent variables. The response variable (dependent variable) we want to investigate is the legibility of a multi-robot system, while the independent variables are the *motion-variables* that affect the behavior of the multi-robot system. In particular, we consider dispersion and stiffness.

For this kind of investigation, it is important to choose the interval of the variables in which we want to recover the function, and the order of the function itself (namely, if we want to approximate with a first or second order model). In general, the area to investigate is chosen after a series of ANOVA studies that follow the optimal direction of the function. However, in this case, the variables can not change over a wide range, because they represent the (virtual) forces that connect the robots: since we want to maintain a physical representation for these forces (in terms of—nonlinear—springs and dampers), the parameters are constrained over a limited set of (physically plausible) values.

In particular, we used the same range that was investigated in [2,3]. Regarding the order of the model, we chose a second order model: this allows us to investigate a model that can approximate also a curvature. In fact, the model we could retrieve from the previous experiments in [3] was only linear (because we investigated only two values for each variable), and we wanted to improve it.

For the second order model, the most used methodology is the center composite design (CCD), which allows to retrieve a model with a reasonable number of experimental points, namely the combinations of the independent variables that must be investigated. The CCD starts from a 2^k factorial plan, where k is number of independent variables, and augments the study with $2k$ axial or star points and a central point. The value of the independent variable corresponding to each point is coded into 5 levels: $+\alpha$, $+1$, 0, $-\alpha$, -1. The additional points are calculated based on the particular characteristics that the study must respect. In this case, we want rotatability to be preserved because it allows to have the same variance of the predicted response for all data points equidistant from the center. To respect this characteristic, we picked the parameter $\alpha = 2^{k/4}$ ($k = 2$ as we have 2 independent variables: dispersion and stiffness). The parameter α defines the distance of the axial points from the center. In this case, the choice of $\alpha = \sqrt{2}$ allows to have also a spherical design, which is a design where all the points belong to a sphere of radius $\sqrt{k}$. Figure 1 reports the CCD that results in 9 different points to investigate.

As we do not have particular restriction on the area of interest, nor a particular knowledge of the model, and we do not need to reduce the number of points, we can use the standard method of CCD and not an optimal design, such as D-Optimal or A-Optimal [12,16].

The experimental plan was replicated in a within-subject study because we need to validate the surface in a general way. Every user performs the full factorial plan in random order to reduce side effects, such as learning effect.

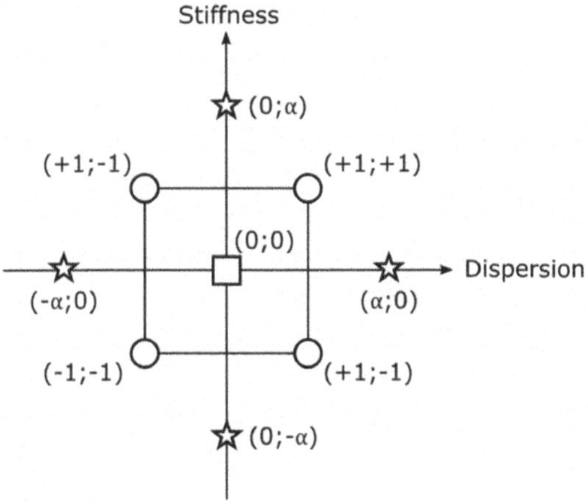

Fig. 1. Central composite design: ○ factorial points, ☆ axial points, □ central point

3.2 Robot Model and Independent Variables

The independent variables considered in this study are dispersion and stiffness, which represent two characteristics of the aggregated behavior of a controlled multi-robot system, composed by N mobile robots. We consider each robot modeled as a double integrator system in an n-dimensional space:

$$M_i \ddot{x}_i = w_i \quad i = 1, \ldots, N \tag{4}$$

where $M_i \in \mathbb{R}^{n\times n}$, positive definite, and $x_i \in \mathbb{R}^n$ are the inertia matrix and the position of the i-th robot, respectively. The control input and all the external forces that act on the robot are included in the term $w_i \in \mathbb{R}^n$. For simplicity of notation, we will hereafter consider $M_i = mI$, where $I \in \mathbb{R}^{n\times n}$ is the identity matrix and m is the mass of the robot.

Dispersion and stiffness are mapped into the control of the robot that is implemented exploiting an artificial potential field technique. This allows us to define the movement and the behavior of the multi-robot system defining only some parameters, two of which are the independent variables. The overall motion of the group is determined by a virtual agent that follows a *minimum-jerk* trajectory $\varphi(t)$. In [7,9], and also in our previous works [2,3], this type of trajectory was found to improve the interaction between humans and robots, and hence the legibility of the system. The virtual agent, which is a fictitious robot, moves from the starting point $\varphi(t) = \varphi(0)$ to the goal $\varphi(t) = \varphi(f)$, with $f > 0$. The 1-dimensional *minimum-jerk* trajectory (5), defined as [7]:

$$\varphi(t) = \varphi(0) + (\varphi(f) - \varphi(0)) \left(10 \left(\frac{t}{f}\right)^3 - 15 \left(\frac{t}{f}\right)^4 + 6 \left(\frac{t}{f}\right)^5\right) \tag{5}$$

is applied componentwise to obtain the desired n-dimensional trajectory.

While the virtual agent follows the *minimum-jerk* trajectory, the multi-robot system is subject to multiple potential fields: attractive and repulsive among the robots to create a cohesive and anti-collision behavior, and an additional attractive potential between the virtual agent and the group of robots.

The potential field acts on the i-th robot by means of the control input ω_i in (4), which can be explicitly rewritten as:

$$M_i \ddot{x}_i = -B_i \dot{x}_i - \nabla V_i \tag{6}$$

where $\dot{x}_i \in \mathbb{R}^n$ is the velocity of the i-th robot and $B_i \in \mathbb{R}^{n \times n}$, positive definite, is the damping matrix, which can be simplified to $B_i = bI$ because we consider a homogeneous friction coefficient (b) along all directions. We introduce damping to obtain a smooth movement, which was shown in [6] to be a factor that influences interaction between human and mobile robots.

The term ∇V_i is the gradient of the potential field V_i, which is composed of three different actions[1]:

$$V_i = \sum_{j=1;j\neq i}^{N} V_{a_{i,j}}(x_i, x_j) + \sum_{j=1;j\neq i}^{N} V_{rep_{i,j}}(x_i, x_j) + V_{v_i}(x_i, x_v) \tag{7}$$

where $V_{a_{i,j}}$ represents the attractive potential between the i-th robot and the j-th robot[2]:

$$V_{a_{i,j}} = \begin{cases} \frac{1}{2} K_{a_{i,j}} (d_{ij} - d_0)^2 & \text{if } d_{ij} \leq d_1 \\ 0 & \text{otherwise} \end{cases} \tag{8}$$

The potential field V_i also prevents the robots from colliding, by means of the repulsive action ($V_{rep_{i,j}}$):

$$V_{rep_{i,j}} = \begin{cases} K_{rep_{i,j}} \left(\frac{1}{3} d_{ij}^3 - d_{min}^3 \ln d_{ij} - \frac{1}{3} d_{min}^3 + d_{min}^3 \ln d_{min}\right) & \text{if } d_{ij} \leq d_{min} \\ 0 & \text{otherwise} \end{cases} \tag{9}$$

[1] When not strictly necessary, we omit the dependence on time of the variables, e.g., $V_i(t) = V_i$, for ease of notation.

[2] $d_{ij} = \|x_i - x_j\|$ is the Euclidean distance between the i-th and the j-th robot.

Then V_{v_i} is the attractive potential between the i-th robot and the virtual agent v (this potential acts only on the real robots, while the virtual agent follows the *minimum-jerk* trajectory without any interference from the robots):

$$V_{v_i} = \begin{cases} \frac{1}{2} K_{v_{i,v}} \left(d_{iv} - d_{0_v} \right)^2 & \text{if } d_{iv} \leq d_{1_v} \\ 0 & \text{otherwise} \end{cases} \tag{10}$$

In the above formulas, the positive constants ($K_{a_{i,j}}$, $K_{v_{i,v}}$ and $K_{rep_{i,j}}$) allow to calibrate the strength of each action. Moreover, the parameters d_1, d_{min} and d_{1_v} define the distance at which the corresponding potential acts, namely they represent the limited range of communication and sensing of the robots.

The independent variables are mapped into:

- Dispersion: d_0;
- Stiffness: $K_{a_{i,j}}$ and $K_{rep_{i,j}}$.

Dispersion is described by the desired distance at which the attractive potential $V_{a_{i,j}}$ keeps the robots. On the other side, stiffness influences the behavior among the robots, namely the power of the connections among them.

As reported in Sect. 3.1, the CCD starts from a factorial plan, which is approximately centered in the area we have already investigated in our previous works [2,3]. The additional points are calculated based on the value of $\alpha = \sqrt{2}$. Table 1 reports the values of the independent variables that are used during the experiments. The other constants that describe the potential fields are kept equal for all the trials, and they are reported in Table 2.

Figure 2 represents some examples of the trajectory of the robots. The positions, previously saved offline, have been then used directly in the Unity program to replicate the same behavior for each experimental trial.

Table 1. Dispersion and stiffness levels.

		$-\alpha$	-1	0	$+1$	$+\alpha$
d_0	[m]	1.378	2	3.5	5	5.621
$K_{a_{i,j}}$	[N/m]	0.017	0.1	0.3	0.5	0.538
$K_{rep_{i,j}}$	[N/m]	8.5	50	150	250	291.5

Table 2. Constant variables.

$K_{v_{i,v}}$	[N/m]	0.4
m	[kg]	10
b	[Ns/m]	1
d_1	[m]	10
d_{1_v}	[m]	10
d_{min}	[m]	0.5
d_{0_v}	[m]	0.1

3.3 Experimental Scenario

In order to retrieve the data for the response surface design, we needed to replicate the experiments for all the users: the most effective, simplest, and fastest way to achieve this was to build a virtual reality environment. According to the results presented in [24], we can state that the results coming from these trials are sufficiently faithful compared to those we could retrieve in a real scenario.

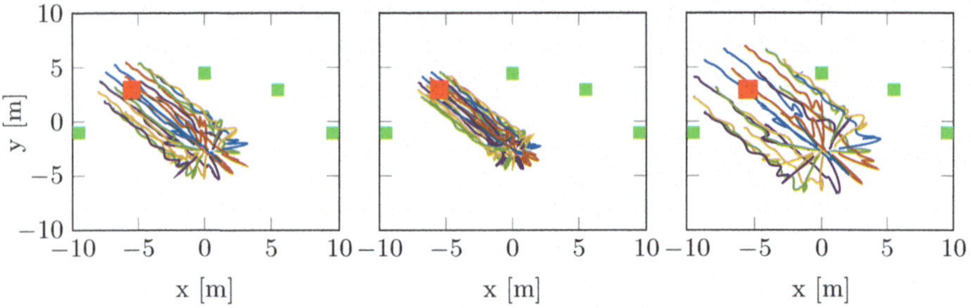

(a) Axial point at $(-\alpha; 0)$. (b) Factorial point at $(+1; -1)$. (c) Central point at $(0; 0)$.

Fig. 2. Robots' trajectory in different point of the CCD, each line represents a robot. The squares represent: *red* the chosen goal and *green* the other goals.

The set-up was a wide area (22×11 m) in which the user shares the space with a group of 20 omnidirectional wheeled robots. The user, in each trial, had to understand where the group of robots was moving towards. The directions were represented by a series of cubes placed in the environment.

The scenario was built with the multi-platform development software Unity, and we used the Oculus Rift to perform the rendering in virtual reality.

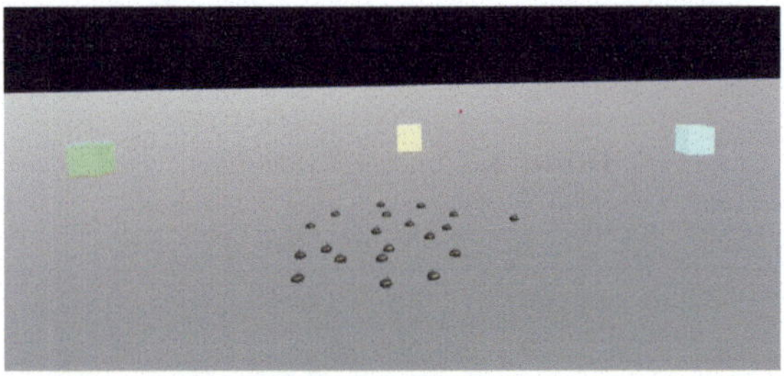

Fig. 3. Virtual reality environment used for the experiments: in dark gray the group of robots, the colored cubes are the possible goals of the group.

3.4 Dependent Variables

Since we want to discover the function that links dispersion and stiffness with the legibility of a multi-robot system, legibility is the dependent variable of our response surface design. Legibility can be decomposed into the correctness of communication and in the time that the communication itself requires. However, in this case, we consider only the response time of each trial because we think that the majority of the users will understand the goal. This belief is supported by the choice of the *minimum-jerk* trajectory, which has been found the most legible in both our previous works [2,3], and, in general, is considered the best choice for this type of communication.

3.5 Users

The CCD was replicated over 18 users (age 26.39 ± 3.3, 5 females and 13 males), according to the within-subjects methodology. They are all volunteers, unrelated to the project, and they are students and researchers of our engineering department. Everyone tested the input method, settled in the virtual reality environment and signed a consent form before the test began. In addition she/he knew that: (1) in each trial the variables that defined the motion of the robots would be changed, (2) the goals were all equally probable, and (3) all the robots moved towards a goal, never changing their destination.

4 Results

Data collected during the experiments were elaborated using Matlab® and its Curve Fitting Tool, with the objective of finding the most appropriate function to represent the data. Several functions can be evaluated, including polynomials and other surfaces: given the design of the CCD experiment, we can consider polynomials of order no larger than two.

Different functions can be compared based on several parameters, that represent how well the function itself fits the available data. In particular, the main parameters that are typically used in this type of statistical analysis are: R-square and the sum of squares due to error (SSE). R-square represents how well the model explain the variation of the experimental data, while SSE depicts the deviation between the data and the fitted values. Hence, a high value of R-square corresponds to a good fit of the data, while a low value of SSE represents a good model for the prediction.

Data were normalized, in order to remove the influence of the different unit of measurement between dispersion and stiffness, namely m and N/m. Furthermore, in order to reduce the effect of the outliers, a robust least-square approach is used for the polynomial fit. In particular, we used the bisquare weights method that gives different weights to the data based on their distance from the fitted model.

Table 3 reports the results of the analysis. The first column of the table contains the type of the considered function: *polyxy* represents a polynomial

Table 3. Summary of the results.

	R-square	SSE
poly00	0.6765	$3.16 \cdot 10^9$
poly01	0.6639	$3.28 \cdot 10^9$
poly10	0.6639	$3.29 \cdot 10^9$
poly11	0.6438	$3.49 \cdot 10^9$
poly21	0.5935	$3.98 \cdot 10^9$
poly12	0.5896	$4.02 \cdot 10^9$
poly22	0.5617	$4.29 \cdot 10^9$
Thin plate	0.034	$9.45 \cdot 10^9$
Biharmonic	0.034	$9.45 \cdot 10^9$

of degree x for the first independent variable, and y for the second one. The corresponding contours are plotted in Fig. 4. In general, the R-square parameter shows that the polynomial functions provide a good approximation of the data. In particular, the best R-square score is obtained for the *poly00*, which is the mean of the value, and it corresponds to a straight plane, parallel to the x-y plane (Fig. 4(a)). However, this function is not useful to represent the data, as it does not highlight any trend in the response, as a result of changes in the independent variables.

Considering the other polynomials, they are all scored with a similar R-square value, and hence can be exploited for representing the acquired data. As can be seen in Fig. 4(c)–(g), all these models provide a higher value of legibility for low values of both dispersion and stiffness: this confirms the preliminary results obtained in [2,3].

The last two methods reported in Table 3 are special methods for surface interpolation. In particular, the thin plate is a polyharmonic spline that allows to calculate the deformation of a thin sheet of metal, while the biharmonc spline are used for minimum curvature interpolation of irregular set of data. None of these methods provide a good R-square score: this was partially expected, since these methods are very application specific. However, we decided to report also these results, for the sake of completeness.

It is worth noting that the SSE level is very large for all the considered functions. This implies that the fitted models are not sufficiently accurate to precisely predict the exact legibility of the system for different values of the independent variables. This was partially expected, due to the large variability introduced by the users. However, all the functions provide similar trends, which, as already mentioned, are in agreement with the preliminary results obtained in [2,3]. Hence, we believe that, while the predicted legibility values are not precise, the trends can be exploited for design purposes: namely, decreasing dispersion and stiffness, in general, leads to better legibility.

Furthermore, the choice of the *minimum-jerk* trajectory led to 159 correct answers over 162 total trials, which confirms the initial hypothesis about the high legibility of this type of trajectory. It justifies also the choice of considering only response time in the analysis of the data.

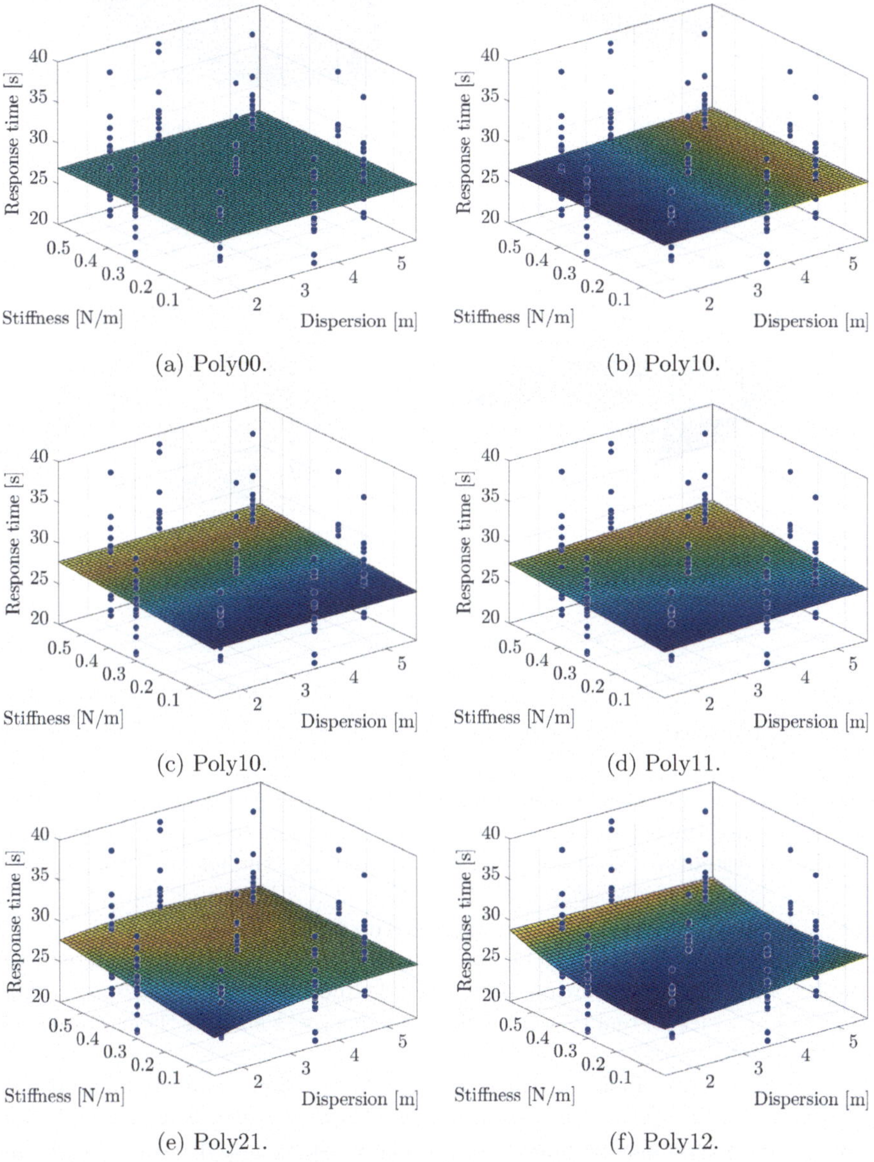

(a) Poly00. (b) Poly10.

(c) Poly10. (d) Poly11.

(e) Poly21. (f) Poly12.

Fig. 4. Model analyzed to fit the data.

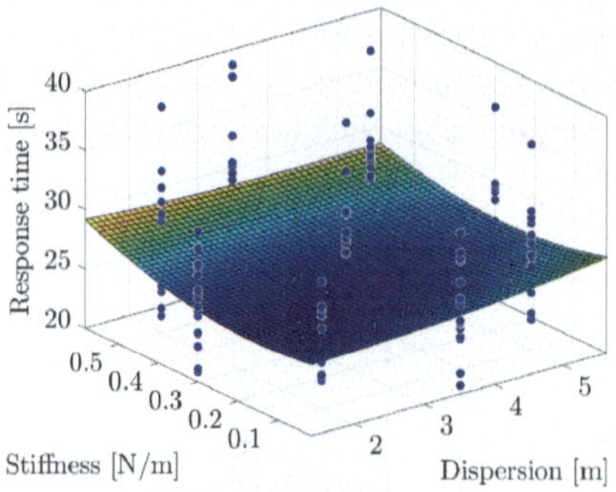

(g) Poly22.

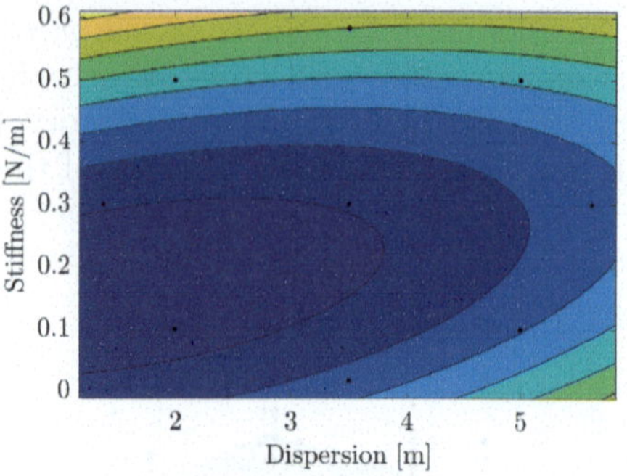

(h) Contour poly22.

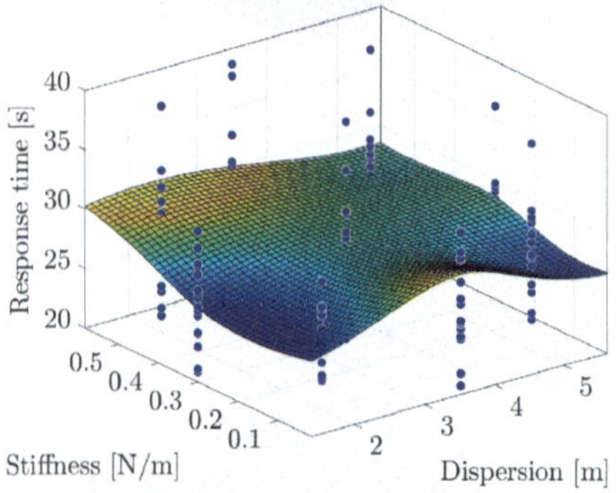

(i) Thin plate.

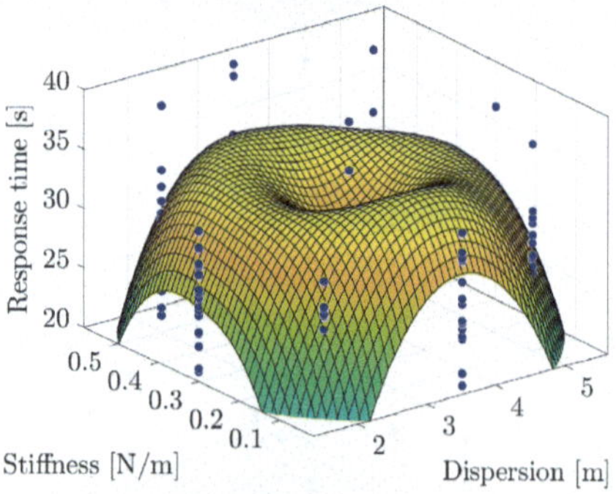

(j) Biharmonic.

Fig. 4. (*continued*)

5 Conclusion

In this paper, we investigated the concept of legibility of a multi-robot system. Considering a group of mobile robots moving in the environment according to an artificial potential based control law, we analyzed the effect of the *motion-variables*, defined as dispersion of the group, and stiffness of the inter-robot interconnections, on the overall legibility of the multi-robot system.

We designed an experimental campaign, where the values of the independent variables were chosen according to the CCD method, and user studies were performed in a virtual reality setup, where the user was sharing the environment with the robots.

While the results did not provide any statistically significant fitting model for the collected data, all the approximation functions provided a clear trend, that confirms the preliminary results of our previous studies [2,3]. We believe that the lack of statistical significance of the fitting model is not surprising, given the large variability introduced by the users themselves.

Future work will aim at improving the results, possibly choosing different *motion-variables*, that better represent the aggregated behavior of the multi-robot system. In particular, we aim at reaching a higher abstraction level, thus possibly leading to more informative quantitative results. Another interesting aspect to investigate is the influence of the number of robots of the group with respect to the legibility.

References

1. Alonso-Mora, J., Lohaus, S.H., Leemann, P., Siegwart, R., Beardsley, P.: Gesture based human-multi-robot swarm interaction and its application to an interactive display. In: Proceedings of 2015 IEEE International Conference on Robotics and Automation (ICRA), pp. 5948–5953. IEEE (2015)
2. Capelli, B., Secchi, C., Sabattini, L.: Communication through motion: legibility of multi-robot systems. In: Proceedings of 2019 International Symposium on Multi-Robot and Multi-Agent Systems (MRS). IEEE (2019)
3. Capelli, B., Villani, V., Secchi, C., Sabattini, L.: Understanding multi-robot systems: on the concept of legibility. In: Proceedings of 2019 IEEE/RSJ International Conference on Intelligent Robots and Systems (IROS). IEEE (2019)
4. Cheung, Y., Chung, J.S.: Cooperative control of a multi-arm system using semi-autonomous telemanipulation and adaptive impedance. In: Proceedings of 2009 International Conference on Advanced Robotics, pp. 1–7, June 2009
5. Cortés, J., Egerstedt, M.: Coordinated control of multi-robot systems: a survey. SICE J. Control Meas. Syst. Integr. **10**(6), 495–503 (2017)
6. Dietz, G., Washington, P., Kim, L.H., Follmer, S., et al.: Human perception of swarm robot motion. In: Proceedings of 2017 CHI Conference on Extended Abstracts on Human Factors in Computing Systems, pp. 2520–2527. ACM (2017)
7. Flash, T., Hogan, N.: The coordination of arm movements: an experimentally confirmed mathematical model. J. Neurosci. **5**(7), 1688–1703 (1985)
8. Franchi, A., Secchi, C., Son, H.I., Bulthoff, H.H., Giordano, P.R.: Bilateral teleoperation of groups of mobile robots with time-varying topology. IEEE Trans. Robot. **28**(5), 1019–1033 (2012)
9. Glasauer, S., Huber, M., Basili, P., Knoll, A., Brandt, T.: Interacting in time and space: investigating human-human and human-robot joint action. In: Proceedings of 19th Annual IEEE International Symposium in Robot and Human Interactive Communication (RO-MAN), pp. 252–257. IEEE (2010)
10. Gromov, B., Gambardella, L.M., Di Caro, G.A.: Wearable multi-modal interface for human multi-robot interaction. In: Proceedings of 2016 IEEE International Symposium on Safety, Security, and Rescue Robotics (SSRR), pp. 240–245. IEEE (2016)
11. Kolling, A., Walker, P., Chakraborty, N., Sycara, K., Lewis, M.: Human interaction with robot swarms: a survey. IEEE Trans. Hum.-Mach. Syst. **46**(1), 9–26 (2016)
12. Kuram, E., Ozcelik, B., Bayramoglu, M., Demirbas, E., Simsek, B.T.: Optimization of cutting fluids and cutting parameters during end milling by using D-optimal design of experiments. J. Cleaner Prod. **42**, 159–166 (2013)
13. Lasota, P.A., Fong, T., Shah, J.A., et al.: A survey of methods for safe human-robot interaction. Found. Trends® Robot. **5**(4), 261–349 (2017)
14. May, A.D., Dondrup, C., Hanheide, M.: Show me your moves! Conveying navigation intention of a mobile robot to humans. In: Proceedings of 2015 European Conference on Mobile Robots (ECMR), pp. 1–6. IEEE (2015)

15. MohaimenianPour, S., Vaughan, R.: Hands and faces, fast: mono-camera user detection robust enough to directly control a UAV in flight. In: Proceedings of 2018 IEEE/RSJ International Conference on Intelligent Robots and Systems (IROS), pp. 5224–5231. IEEE (2018)
16. Montgomery, D.C.: Design and Analysis of Experiments. Wiley, Hoboken (2017)
17. Nagi, J., Giusti, A., Gambardella, L.M., Di Caro, G.A.: Human-swarm interaction using spatial gestures. In: Proceedings of 2014 IEEE/RSJ International Conference on Intelligent Robots and Systems (IROS), pp. 3834–3841. IEEE (2014)
18. Palafox, O., Spong, M.: Bilateral teleoperation of a formation of nonholonomic mobile robots under constant time delay. In: Proceeding of 2009 IEEE/RSJ International Conference on Intelligent Robots and Systems (IROS), pp. 2821–2826, October 2009
19. Rodríguez-Seda, E.J., Troy, J.J., Erignac, C.A., Murray, P., Stipanović, D.M., Spong, M.W.: Bilateral teleoperation of multiple mobile agents: coordinated motion and collision avoidance. IEEE Trans. Control Syst. Technol. **18**(4), 984–992 (2010)
20. Sabattini, L., Secchi, C., Capelli, B., Fantuzzi, C.: Passivity preserving force scaling for enhanced teleoperation of multirobot systems. IEEE Robot. Autom. Lett. **3**(3), 1925–1932 (2018)
21. St Clair, A., Mataric, M.: How robot verbal feedback can improve team performance in human-robot task collaborations. In: Proceedings of 10th Annual ACM/IEEE International Conference on Human-Robot Interaction (HRI), pp. 213–220. ACM (2015)
22. Szafir, D., Mutlu, B., Fong, T.: Communication of intent in assistive free flyers. In: Proceedings of 9th Annual ACM/IEEE International Conference on Human-Robot Interaction (HRI), pp. 358–365. ACM (2014)
23. Szafir, D., Mutlu, B., Fong, T.: Communicating directionality in flying robots. In: Proceedings of 10th Annual ACM/IEEE International Conference on Human-Robot Interaction (HRI), pp. 19–26. ACM (2015)
24. Villani, V., Capelli, B., Sabattini, L.: Use of virtual reality for the evaluation of human-robot interaction systems in complex scenarios. In: Proceedings of 27th IEEE International Symposium on Robot and Human Interactive Communication (RO-MAN), pp. 422–427. IEEE (2018)
25. Villani, V., Sabattini, L., Riggio, G., Secchi, C., Minelli, M., Fantuzzi, C.: A natural infrastructure-less human-robot interaction system. IEEE Robot. Autom. Lett. **2**(3), 1640–1647 (2017)
26. Villani, V., Sabattini, L., Secchi, C., Fantuzzi, C.: Natural interaction based on affective robotics for multi-robot systems. In: Proceedings of the IEEE International Symposium on Multi-Robot and Multi-Agent Systems (MRS), Los Angeles, CA, USA, pp. 56 – 62, December 2017

Closing the Feedback Loop: The Relationship Between Input and Output Modalities in Human-Robot Interactions

Tamara Markovich(✉), Shanee Honig, and Tal Oron-Gilad

Ben-Gurion University of the Negev, Beer-Sheva, Israel
{tamarama,shonig}@post.bgu.ac.il, orontal@bgu.ac.il

Abstract. Previous studies suggested that communication modalities used for human control and robot feedback influence human-robot interactions. However, they generally tended to focus on one part of the communication, ignoring the relationship between control and feedback modalities. We aim to understand whether the relationship between a user's control modality and a robot's feedback modality influences the quality of the interaction and if so, find the most compatible pairings. In a laboratory Wizard-of-Oz experiment, participants were asked to guide a robot through a maze by using either hand gestures or vocal commands. The robot provided vocal or motion feedback to the users across the experimental conditions forming different combinations of control-feedback modalities. We found that the combinations of control-feedback modalities affected the quality of human-robot interaction (subjective experience and efficiency) in different ways. Participants showed less worry and were slower when they communicated with the robot by voice and received vocal feedback, compared to gestural control and receiving vocal feedback. In addition, they felt more distress and were faster when they communicated with the robot by gestures and received motion feedback compared to vocal control and motion feedback. We also found that providing feedback improves the quality of human-robot interaction. In this paper we detail the procedure and results of this experiment.

Keywords: Human-robot interaction · Feedback loop · Navigation task · Feedback by motion cues · Stimulus-response compatibility

1 Introduction

The "Feedback loop" is an important feature of interactive systems. It represents the nature of the interaction between a person and a dynamic system: the user provides input to the system in order to achieve a goal, gets output (feedback) from the system and interprets it. This interpretation affects the user's next action, beginning the cycle again [1]. Robots interacting with humans should be able to react to commands given by the users as well as provide feedback. Some studies emphasize the importance of giving feedback by robots to humans during the interaction [2, 3]. The feedback can be provided in different modalities—using tactile devices, verbal feedback, and visual feedback through screens or lights and more. Similarly, humans can communicate with

F. Ferraguti et al. (Eds.): HFR 2019, SPAR 12, pp. 29–42, 2020.
https://doi.org/10.1007/978-3-030-42026-0_3

the robot through several modalities, including gestures, voice commands and touch screens. In the literature, different communication modalities have been investigated for the robot's feedback or the human's control showing that communication modalities influence human-robot interactions [4, 5]. However, these investigations provide an incomplete understanding of the interaction since they generally focus only on one part of the communication (from the robot to the human or vice versa), ignoring the relationship between control and feedback modalities.

According to Greenwald [6, 7], there are stimuli modalities that most compatibly map to certain response modalities. This statement is an extension of a principle called Stimulus-Response (S-R) compatibility. According to this principle, when the relation between displays and controls, or stimuli and responses, is direct and natural, the responses are faster and more accurate than when they are incompatible [8]. According to Greenwald's theory, a stimulus in a certain modality would activate the sensory images in this modality that in turn would initiate the response that produces this sensory effect. Under this hypothesis, there is compatibility between stimulus and response that would produce shorter Reaction Times (RTs). This hypothesis was tested in many experiments in which auditory and visual stimulus modalities were paired with spoken or manual responses. The results showed, as predicted, an interaction effect between stimulus and response modalities: compatible combinations (visual-manual, auditory-spoken) had shorter RT than incompatible combinations. Greenwald's suggestion is based on the ideomotor theory, which suggests that actions are initiated by an anticipation of their sensory effects. For example, an image of a word's sound primes speech of the word and visual image of the written word primes writing. Ideomotor interpretation of Greenwald's results is that compatible combinations allow easier encoding because the proper response is selected without the need to translate the stimulus code into another modality. Wickens et al. [9] expanded the concept of stimulus-response compatibility to incorporate a mediating central processing (C) component (S-C-R): operators incorporate the stimulus information into a mental model of the system and then choose a response action. In their experiment, they compared task performance between combinations of auditory and visual stimulus modalities and speech and manual response modalities for spatial or verbal task. They showed that there is an association between a task's processing code (verbal or spatial) and stimulus-response modalities: for the verbal task, RT was shortest with the combination of auditory input and speech output. For the spatial task, RT was shortest with the combination of visual input and manual output.

Applying these ideas to human-robot interaction, it is possible that there are compatibility effects between control modalities and feedback modalities that may affect the quality of the human-robot interaction by influencing the speed and accuracy of human response, as well as the subjective experience of the human during the interaction. To evaluate how the relationship between control modalities and feedback modalities may impact the human-robot interaction, we planned an experiment in which participants were asked to perform a navigation task (guiding a robot through a maze) while changing the modality of the commands (hand gestures vs. voice commands) and the robot's feedback (vocal feedback, motion feedback or without feedback). The specific commands and feedback used in the experiment were selected based on the results of a preliminary experiment, which evaluated which gestures or spoken commands people are more likely to use or prefer to use when navigating the robot through a maze, and what feedback the robot should give in order to facilitate clear and intuitive communication.

2 Goals and Hypotheses

We aim to understand whether the mapping of human control modality to robot feedback modality during the interaction influences the quality of the interaction and if so, find the most compatible mappings. Our main hypothesis, based on Ideomotor theory and S-R compatibility effects, is that there is a compatibility effect between control modalities (response) and feedback modalities (stimulus). Moreover, we expect to find that vocal control would be most compatibly mapped to feedback in the same modality and gestural control will be most compatibly mapped to motion feedback. Vocal stimulus and vocal response combination were found to be compatible by Greenwald since the sensory effects of vocal response and stimulus are the same. Similarly, we argue that the sensory effect of gestural action is motion, so motion feedback would initiate the gestural action.

Based on previous literature which showed the importance of providing feedback in human-robot interactions [2, 3], we hypothesize that the subjective experience and task performance would be better when the robot provides feedback than when the robot doesn't provide feedback. Lastly, we mentioned Wickens' experiments that demonstrated the efficacy of a combination of manual control and visual feedback for spatial tasks [9]. Therefore, we hypothesize that the efficacy of different combinations of control-feedback would be dependent on the type of task. Specifically, we expect that for a spatial task like maze navigation, the combination of gestural control (that can be considered as a type of manual response) and motion feedback (that can be considered as a type of visual feedback) would produce a more efficient interaction (lower RT and better subjective experience) than the combination of vocal control and vocal feedback.

3 Method

3.1 Overview

A WoZ experiment was conducted, using a within-subjects factorial design. The first factor was control modality, divided into two options: vocal control or gestural control. The second factor was feedback modality, divided into three alternatives: motion feedback, vocal feedback, and no feedback. Participants performed a navigation task, guiding a robot out of a maze by giving it navigation commands. The modality of the commands and the robot's feedback changed across the conditions of the experiment. Participants gave the robot four successive commands before they ordered it to start moving. The robot executed these commands consecutively once it received the order to start moving. When the robot had finished moving, the participant gave it a new set of four commands and so on. The maze had seven different entrance and exit points. Each participant performed the task seven times, each time using a different route and different combinations of control-feedback modalities.

3.2 Participants

Twenty-three students (12 female, 11 male), aged 21–27, from Ben-Gurion University of the Negev participated in this study. In compensation, they received extra course credit.

3.3 Robot

The experiment used a "Burger" variant of TurtleBot3 (see Fig. 1), running ROS. The Burger variant has two wheels driven by servos and a ball bearing to keep its balance. Using the servos, the robot can go forward, backward, and turn around itself. The robot was teleoperated from a remote PC (Asus ZenBook UX410UA-GV349T laptop). A Bluetooth speaker was connected to the robot to enable it to provide vocal feedback.

Fig. 1. TurtleBot3 burger

3.4 Experimental Design

Maze Design. The maze used in the experiment was marked on the floor using masking tape. We put corresponding numbers on each route's entrance and exit point to clarify where each route begins and ends (see Fig. 2). To construct the maze in such a way that all routes would not differ in their complexity, we used the choice-clue wayfinding model by Raubal and Egenhofer [10]. According to this model, the complexity of wayfinding task is determined by counting the decision points where people have more than one option to continue the task. Thus, in the maze we built, all routes had the same number of decision points with more than one option (four decision points in each route). A run was defined as a navigation of the robot from a beginning point to an end point of a certain path (e.g., from 5 to 5).

Control and Feedback. The set of commands, described in Table 1, was used to guide the robot to turn right, turn left, go forward and start moving. Demonstration of possible gestures commands can be seen in Fig. 3. An order to turn, made the robot turn ninety degrees to the right or left, respectively. An order to go forward advanced the robot for a constant distance of three tiles. The possible feedbacks options, displayed in Table 2, were: providing the participant information about the robot's understanding (i.e., if the

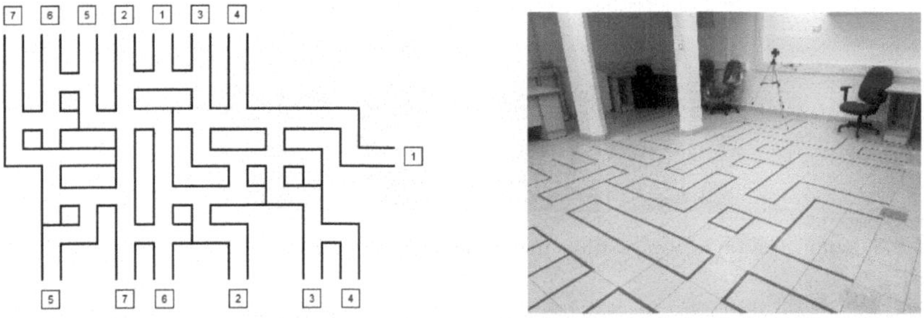

Fig. 2. The maze. Left. A schematic representation of the maze. Each path starts and ends with the same number. Right. A pictorial view of the maze markings on the floor

command was understood or not), about its intention to start moving or about its ability to execute the command. The robot's understanding or misunderstanding feedback appeared right after the participants gave the robot an order to turn right, turn left or go forward. The likelihood for command to be understood or not was set randomly (with 0.9 probability for the appearance of "understood" feedback and 0.1 for "not understood" feedback). If a certain command was not understood, the participant repeated it. In the no-feedback conditions, if a certain command was not understood, after receiving a command to start moving, the robot halted instead of executing the commands and the participant was asked to repeat the last sequence of four commands. The feedback about the robot's intention to start moving appeared at the end of sequence of the four commands, after the participant ordered the robot to start moving. The feedback about the robot's ability to execute the command appeared during the execution of the command and not immediately after the participant gave the problematic command. For example, if the participant asked the robot to perform an action that would cause the robot to clash into a maze wall, during the execution, the robot halted and informed the participant that the command could not be executed. In such a case, the participant gave the robot a new set of four commands.

Table 1. Set of commands for the maze navigation task, the vocal commands used by the users in the vocal control condition and the gestures they used in gestural control condition.

Command	Vocal commands	Gestures
Turn right	Turn right	Pointing with the thumb to the right
Turn left	Turn left	Pointing with the thumb to the left
Go forward	Go forward	A tense arm rises from the bottom up
Start moving	Start	Hands clinging

One experimenter was present during the study, responsible for instructing the participants on the maze task, the experimental procedure and running the study. Using a keyboard, the experimenter controlled the robot's movements, making it appear as if the robot is responding to the participant's commands and providing motion feedback in

Fig. 3. A demonstration of the gestures the participants were asked to use in the gestural control condition

Table 2. Set of feedback messages that the robot provided for the maze navigation task, the vocal feedback was provided by the robot in the vocal feedback condition and the motion feedback in the motion feedback condition.

Feedback message	Vocal feedback	Motion feedback
The command was understood	Understood	Move one step forward and backward
The command was not understood	Did not understand	Turn to the right, then to the left and return
The command cannot be executed	Cannot execute	Turn a little to left and return
Start moving	Starting to move	Move a little forward

the relevant conditions. In the vocal feedback condition, the experimenter played pre-recorded vocal feedback recordings using R-studio. Since the motion feedback duration was longer than the recorded vocal feedback messages, we played three "beeps" using the "beepr" package before the vocal feedback. By this the length in time of the two feedback modalities was similar and we could eliminate feedback duration as a possible confounding variable.

Subjective Measurement. To measure the subjective experience of the participants, at the end of each run participants filled out two online surveys that were administered via Google Form:

1. The System Usability Scale (SUS) [11]: A ten-item scale that gives a global view of subjective assessments of usability. SUS yields a single number (in range 0–100) representing a composite measure of the overall usability of the system. A system that receives a SUS score of 68 or above is considered usable [12]. We adjusted the scale to our goal in such a way that the questions referred to the interaction with the robot. For example, instead of the item "I felt very confident using the system", we used "I felt very confident communicating with the robot". All items were measured on a five-point scale ranging from one (strongly disagree) to five (strongly agree). The SUS score was calculated for each run.
2. Stress State Questionnaire (DSSQ) [13]: Comprehensive assessment of subjective states in performance contexts, based on a factor model that differentiates 11 primary state factors, which cohere around three higher-order dimensions of task engagement,

distress, and worry. We used twenty relevant items. All items were measured on a five-point scale ranging from zero (strongly disagree) to four (strongly agree). We calculated the participant's engagement, distress and worry scores for each run. The possible worry score is in a range of 0–24, while distress and engagement scores are in a range of 0–28.

3.5 Procedure

The experiment began with a collection of participants demographic data and consent forms. First, participants were instructed about the maze navigation task and introduced to the possible commands they could use and how to communicate them to the robot by gestures or voice commands. To train, they were asked to navigate the robot from one point to another twice, one time using voice commands and the second time using gestures. After training, participants performed the task for seven runs. At the end of each run, they were asked to complete the SUS and DSSQ. By the end of the last run, they were given a post-experiment questionnaire that included questions about their preferences for communication modalities combinations. In each run, except the last one, participants were instructed on which modality they should use. In addition, we told them if they should expect the robot to provide feedback and if so the type of feedback to expect. In each run, there was a different combination of control-feedback modalities. The order of combinations was counterbalanced among participants. In the last run, participants could freely choose the way they wanted to communicate with the robot as well as their preferred feedback modality (or without feedback). All runs were filmed by two video cameras for post-experiment analysis.

3.6 Measures

For each participant, we measured *efficiency* and subjective experience. To rate the efficiency, we measured the participant's reaction time (RT): the average time from the moment the robot produced feedback till the following command the participant gave to the robot. We also measured *accuracy* by counting the number of errors participants made in performing the task (whether the subject led the robot on a longer path than the shortest possible path). We evaluated *subjective experience* using the custom SUS and DSSQ surveys that participants were asked to fill.

3.7 Analysis

To analyze the recorded videos, we used Observer XT, professional software for the collection, analysis, and presentation of videos [14]. By that, we extracted variables that are related to task performance, such as duration of the run, participants reaction times and accuracy. The statistical analysis was done in SPSS. We used a generalized estimating equation model framework (GEE) including the fixed effects and one random effect which accounted for individual differences among participants.

Subjective Experience. The DSSQ consists of three factors: distress, worry, and task engagement. The SUS yields one factor - SUS score. We used GEE framework to calculate the effects of control modality, feedback modality and run duration (the time from the beginning of the navigating task until the robot reached the end exit of the maze) on distress, worry, engagement and SUS score. The second order interaction control modality * feedback modality was included in the model.

Efficiency. For RT analysis, we log-transformed RT, to lnRT, which has a normal distribution. We used GEE framework to calculate the effects of control modality and feedback modality (including the two-way interaction) on lnRT.

Accuracy. We evaluated if the participant made navigation errors during a run using a binary variable. When the participant navigated the robot in the shortest path, without deviations, the Accuracy factor was marked as 1. For all other cases, i.e., when the participant did not navigate the robot in the shortest path it was marked as 0.

4 Results

4.1 Subjective Experience Analysis

Distress. Tests of model effects of distress are summarized in Table 3. Feedback modality had significantly contributed to distress ($Chi^2(2) = 12.393$, $p = 0.002$), as did the control modality and feedback modality interaction ($Chi^2(2) = 6.809$, $p = 0.033$). As can be seen in Fig. 4., when there was no feedback, distress was higher ($M = 5.63$, $SD = 0.645$) than in the cases that included motion feedback ($M = 4.92$, $SD = 0.636$) or vocal feedback ($M = 3.80$, $SD = 0.664$). When the participants received motion feedback, they felt higher distress if they communicated with the robot by gestures ($M = 5.56$, $SD = 0.727$) than by voice ($M = 4.28$, $SD = 0.706$). Run duration also significantly contributed to distress ($Chi^2\ (1) = 9.388$, $p = 0.002$). Increase in distress was correlated with longer runs (i.e., longer interaction duration).

Table 3. Tests of model effects for distress

Source	Wald Chi-Square	df	Sig.
(Intercept)	1.971	1	.160
Feedback modality	12.393	2	.002
Control modality	.224	1	.636
Run duration	9.388	1	.002
Control modality * Feedback modality	6.809	2	.033

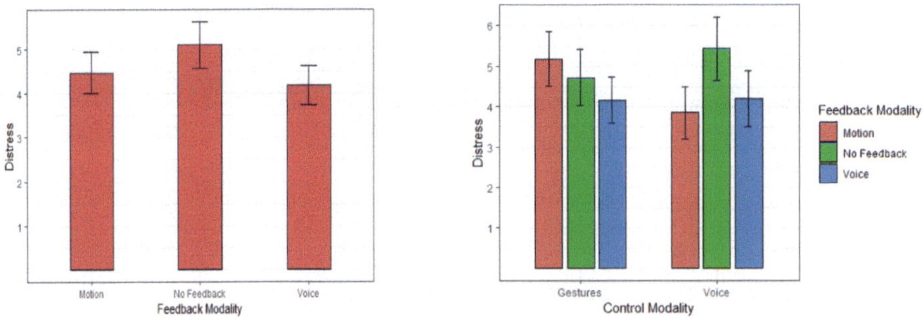

Fig. 4. Distress by feedback modality (left) and distress by the control and feedback modality interaction (right)

Worry. Tests of model effects of worry are summarized in Table 4. The interaction between control modality and feedback modality had significantly contributed to worry ($Chi^2(2) = 10.866$, $p = 0.004$). Figure 5 shows the interaction effect; when there was vocal feedback, there was higher worry if the participants communicated with the robot by gestures ($M = 6.90$, $SD = 0.877$) rather than by voice ($M = 5.29$, $SD = 0.887$).

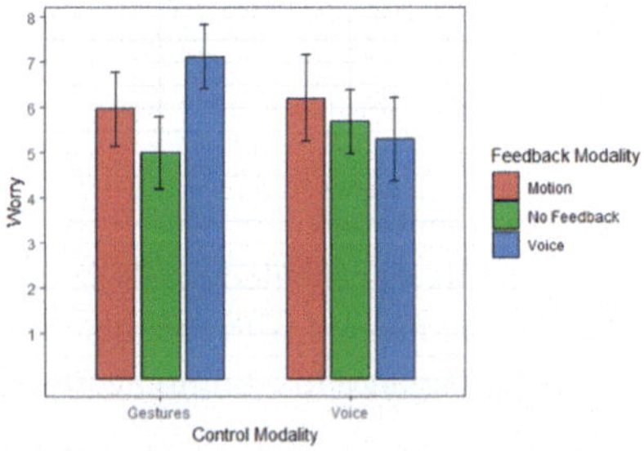

Fig. 5. Worry by control and feedback modality

Table 4. Tests of model effects for worry

Source	Wald Chi-Square	df	Sig.
(Intercept)	15.899	1	.000
Feedback modality	1.222	2	.543
Control modality	.447	1	.504
Run duration	.692	1	.405
Control modality * Feedback modality	10.866	2	.004

Engagement. Tests of model effects of engagement are summarized in Table 5. Feedback modality had significantly contributed to task engagement ($Chi^2(2) = 6.113$, $p = 0.047$). Higher task engagement was seen in motion feedback condition ($M = 20.68$, $SD = 0.810$) compared to no feedback ($M = 20.60$, $SD = 0.823$) or vocal feedback condition ($M = 19.74$, $SD = 0.850$).

Table 5. Tests of model effects for engagement

Source	Wald Chi-Square	df	Sig.
(Intercept)	304.779	1	.000
Feedback modality	6.113	2	.047
Control modality	.010	1	.919
Run duration	.132	1	.716
Control modality * Feedback modality	2.326	2	.313

SUS Score. Tests of model effects of SUS score are summarized in Table 6.

Table 6. Tests of model effects for SUS score

	Wald Chi-Square	df	Sig.
(Intercept)	393.751	1	.000
Control modality	7.821	1	.005
Feedback modality	1.598	2	.450
Run duration	3.225	1	.073
Control modality * Feedback modality	1.217	2	.544

Control modality had significantly contributed to SUS score ($Chi^2(1) = 7.821$, $p = 0.005$). When the participants communicated with the robot by voice, SUS scores were higher (M = 78.50, SD = 2.488) compared to communication by gestures (M = 71.90, SD = 2.441).

4.2 Efficiency Analysis

RT. Tests of model effects for lnRT are summarized in Table 7.

Table 7. Tests of model effects for lnRT

	Wald Chi-Square	df	Sig.
(Intercept)	36.728	1	.000
Feedback modality	2.146	1	.143
Control modality	9.952	1	.002
Control modality * Feedback modality	.402	1	.526

Control modality significantly contributed to lnRT ($Chi^2(1) = 9.952, p = 0.002$). Participants were slower to respond in the vocal control condition, compared to the gestural control condition. The interaction between control modality and feedback modality had not significantly contributed to lnRT. However, when looking at pairwise comparisons of estimated marginal means of lnRT, we found some significant means differences. The means of the group that used vocal control and received motion feedback (VM) and the group that used gestural control and received motion feedback (GM) were significantly different ($p = 0.001$). In addition, the mean of the group that used vocal control and received vocal feedback (VV) was significantly different ($p = 0.001$) from the mean of the group that used gestural control and received vocal feedback (GV). The means of RT for these groups are demonstrated in Fig. 6. The VM group were slower ($M = 0.82$, $SD = 0.048$) than GM group ($M = 0.65$, $SD = 0.021$) and VV group were slower ($M = 0.73$, $SD = 0.025$) than GV group ($M = 0.63$, $SD = 0.043$).

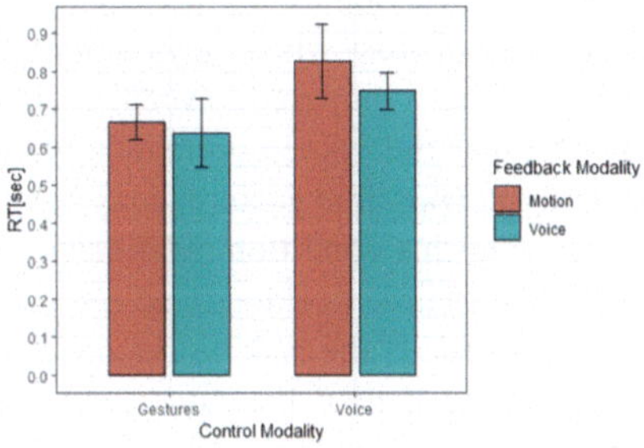

Fig. 6. RT by control and feedback modality

Accuracy. Overall, accuracy rates were quite high ($M = 0.83$, $SD = 0.374$), meaning that participants erred only in about 13% of the runs. Tests of model effects of accuracy are summarized in Table 8. Feedback modality had significantly contributed to accuracy ($Chi^2(2) = 6.428$, $p = 0.04$). Participants erred more during the no feedback condition ($M = 0.72$, $SD = 0.061$), compared to the vocal feedback condition ($M = 0.89$, $SD = 0.04$) and motion feedback condition ($M = 0.90$, $SD = 0.039$).

Table 8. Tests of model effects for accuracy

	Wald Chi-Square	df	Sig.
(Intercept)	979.287	1	.000
Control modality	.053	1	.818
Feedback modality	6.428	2	.040
Control modality * Feedback modality	2.371	2	.306

5 Discussion

The aim of this experiment was to evaluate how the relationship between control modalities and feedback modalities impact the human-robot interaction. Participants were asked to lead the robot to perform a maze navigation task while changing the modality of the commands and the robot's feedback across the different conditions of the experiment. We hypothesized that when the robot provided feedback, the subjective experience and task performance would be better than in the case when the robot didn't provide feedback. The results support this hypothesis, showing that participants felt more distress and were less accurate when the robot didn't provide feedback. We also expected to find that vocal control would be most compatibly mapped to vocal feedback and gestural control would be most compatibly mapped to motion feedback. Some findings support this hypothesis. The worry levels were low generally, but participants showed less worry when they communicated with the robot by voice and received vocal feedback, compared to gestural control and receiving vocal feedback. However, contrary to our expectations, participants felt more distress when they communicated with the robot by gestures and received motion feedback compared to vocal control and motion feedback. A possible explanation may be that the vocal commands were more intuitive and/or easier to remember. Memorizing the possible gestures may have been harder [15], leading to higher cognitive load during this condition. Matthews et al. [13, 16] suggested that the distress response is linked to task workload and a sense of being overwhelmed by task demands is central to distress. If the task workload was greater in the gestural control condition, it may explain the higher distress over the vocal control condition. However, as we expected, participants were faster when they received motion feedback and communicated with the robot by gestures compared to when they communicated by voice. When the participants communicated with the robot by voice and received vocal feedback, they were slower than in the case where they used gestures and received vocal feedback. This finding is also not consistent with our hypothesis since we expected to find that vocal control would be most compatibly mapped to feedback in the same modality. A possible explanation could be that the efficacy of different combinations of control-feedback would be dependent on the task type, so that gestural control and motion feedback would produce a more efficient interaction when performing a spatial task like navigation. Partially confirming this hypothesis, we found that participants were faster when they controlled the robot by gestures rather than by voice and showed more task engagement when they received motion feedback in comparison to no feedback or vocal feedback. It seems that although the objective measurement (reaction time) supported the idea that gestural control is more effective when the task is spatial, the subjective experience shows the opposite. Even though both control modalities received high SUS scores (above 70) that indicate high usability [12], the usability of the robotic system was perceived to be higher when the participants communicated with the robot by voice and not by gestures. Nielsen [17] described usability by five attributes: learnability, efficiency, memorability, low error rate or easy error recovery, and satisfaction. As mentioned by Bau and Mackay [15], gesture-based interaction is ultimately more efficient for experts, while novices need extra support to learn. Since it may be have been harder for our participants, that are novice in gesture-based interaction, to learn and remember the possible gestures compared to the vocal commands, this may be the factor that made participants rate gestural communication with lower usability than vocal communication. In addition, perhaps they found it unusual to gesture without a human dialog partner and unnecessary since the system understood them via voice input, as was found in Beringer research [18]. Consistently, in the last trial of the experiment in which participants could freely choose the way they wanted to communicate with the robot as well as their preferred feedback modality, most of the participants chose vocal control without feedback (9 out of 23), while the preferences of the rest of participants was divided relatively equal among the other possible control-modality combinations. It makes sense that participants preferred the control modality that they perceived as more usable.

In conclusion, the aim of this experiment was to evaluate whether and how the relationship between control and feedback modalities impact the human-robot interaction. We found that different combinations of control-feedback modalities affected the quality of the interaction in different ways. We also found that providing feedback improved the quality of the interaction. Our findings highlight the importance of investigating the relationship between control and feedback modalities for improving the quality of human-robot interaction, instead of focusing only on one part of the communication. Our experiment has some limitations that may have affected the results that were not conclusive regarding the question of how the mapping of human control modality to robot feedback modality influences the quality of the interaction. In our experiment, the feedback that participants received did not include information that was required for task completion. Future research should examine this question using a task in which the feedback provided by the robot is critical to task success. In addition, it would be beneficial to evaluate a verbal task to see if the type of task has influence when exploring the control and feedback modalities relationship impact on the human-robot interaction.

Acknowledgments. This research was supported by the Helmsley Charitable Trust through the Agricultural, Biological and Cognitive Robotics Center at Ben-Gurion University of the Negev. The second author, SH is also supported by Ben-Gurion University of the Negev through the High-tech, Bio-tech and Chemo-tech Scholarship.

References

1. Dubberly, H., Pangaro, P., Haque, U.: ON MODELING What is interaction?: are there different types? Interactions **16**(1), 69–75 (2009)
2. Mirnig, N., Riegler, S., Weiss, A., Tscheligi, M.: A case study on the effect of feedback on itinerary requests in human-robot interaction. In: 2011 IEEE RO-MAN, pp. 343–349. IEEE, July 2011
3. Mohammad, Y., Nishida, T.: Interaction between untrained users and a miniature robot in a collaborative navigation controlled experiment. Int. J. Inf. Acquis. **5**(04), 291–307 (2008)
4. Redden, E.S., Carstens, C.B., Pettitt, R.A.: Intuitive speech-based robotic control (No. ARL-TR-5175). Army Research Lab Aberdeen Proving Ground MD Human Research and Engineering Directorate (2010)
5. Perrin, X., Chavarriaga, R., Ray, C., Siegwart, R., Millán, J.D.R.: A comparative psychophysical and EEG study of different feedback modalities for HRI. In: Proceedings of the 3rd ACM/IEEE International Conference on Human Robot Interaction, pp. 41–48. ACM, March 2008

6. Greenwald, A.G.: A choice reaction time test of ideomotor theory. J. Exp. Psychol. **86**, 20–25 (1970)
7. Greenwald, A.G.: Sensory feedback mechanisms in performance control: with special reference to the ideo-motor mechanism. Psychol. Rev. **77**(2), 73 (1970)
8. Proctor, R.W., Vu, K.P.L.: Stimulus-Response Compatibility Principles: Data, Theory, and Application. CRC Press, Boca Raton (2006)
9. Wickens, C.D., Sandry, D.L., Vidulich, M.: Compatibility and resource competition between modalities of input, central processing, and output. Hum. Factors **25**(2), 227–248 (1983)
10. Raubal, M., Egenhofer, M.J.: Comparing the complexity of wayfinding tasks in built environments. Environ. Plan. **25**(6), 895–913 (1998)
11. Brooke, J.: SUS-a quick and dirty usability scale. Usability Eval. Ind. **189**(194), 4–7 (1996)
12. Bangor, A., Kortum, P., Miller, J.: Determining what individual SUS scores mean: adding an adjective rating scale. J. Usability Stud. **4**(3), 114–123 (2009)
13. Matthews, G., Campbell, S.E., Falconer, S., Joyner, L.A., Huggins, J., Gilliland, K., Grier, R., Warm, J.S.: Fundamental dimensions of subjective state in performance settings: task engagement, distress, and worry. Emotion **2**(4), 315 (2002)
14. Noldus, L.P., Trienes, R.J., Hendriksen, A.H., Jansen, H., Jansen, R.G.: The Observer Video-Pro: new software for the collection, management, and presentation of time-structured data from videotapes and digital media files. Behav. Res. Methods Instrum. Comput. **32**(1), 197–206 (2000)
15. Bau, O., Mackay, W.E.: OctoPocus: a dynamic guide for learning gesture-based command sets. In: Proceedings of the 21st Annual ACM Symposium on User Interface Software and Technology, pp. 37–46. ACM, October 2008
16. Matthews, G., Emo, A.K., Funke, G., Zeidner, M., Roberts, R.D., Costa Jr., P.T., Schulze, R.: Emotional intelligence, personality, and task-induced stress. J. Exp. Psychol. Appl. **12**(2), 96 (2006)
17. Nielsen, J.: Usability Engineering. Academic Press, Cambridge (1993)
18. Beringer, N.: Evoking Gestures in SmartKom – Design of the Graphical User Interface (2001)

Incremental Motion Reshaping of Autonomous Dynamical Systems

Matteo Saveriano[1(✉)] and Dongheui Lee[2,3]

[1] University of Innsbruck, Innsbruck, Austria
matteo.saveriano@uibk.ac.at
[2] Technical University of Munich, Munich, Germany
[3] German Aerospace Center (DLR), Weßling, Germany

Abstract. This paper presents an approach to incrementally learn a reshaping term that modifies the trajectories of an autonomous dynamical system without affecting its stability properties. The reshaping term is considered as an additive control input and it is incrementally learned from human demonstrations using Gaussian process regression. We propose a novel parametrization of this control input that preserves the time-independence and the stability of the reshaped system, as analytically proved in the performed Lyapunov stability analysis. The effectiveness of the proposed approach is demonstrated with simulations and experiments on a real robot.

Keywords: Incremental learning of stable motions · Dynamical systems for motion planning

1 Introduction

In unstructured environments, the successful execution of a task may depend on the capability of the robot to rapidly adapt its behavior to the changing scenario. Behavior adaptation can be driven by the robot's past experience or, as in the Programming by Demonstration (PbD) paradigm [20], by a human instructor that shows to the robot how to perform the task [16,19,21]. Demonstrated skills can be encoded in several ways. Dynamical systems (DS) are a promising approach to represent demonstrated skills and plan robotic motions in real-time. DS have been successfully used in a variety of robotic applications including point-to-point motion planning [3,6,14,17], reactive motion replanning [1,2,5,7], and learning impedance behaviors from demonstrations [18,24].

Although DS are widely used in robotics applications, there is not much work on incremental learning approaches for DS. Some approaches have extended the Dynamic Movement Primitives (DMPs) [6] towards incremental learning. In [8], authors provide a corrective demonstration to modify a part of a DMP trajectory

M. Saveriano—This work was carried out when the author was at the Human-centered Assistive Robotics, Technical University of Munich.

F. Ferraguti et al. (Eds.): HFR 2019, SPAR 12, pp. 43–57, 2020.
https://doi.org/10.1007/978-3-030-42026-0_4

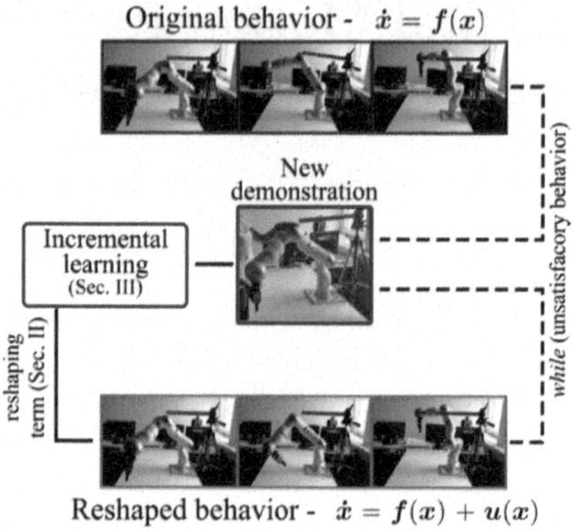

Fig. 1. System overview. (Top) The user observes the original robot behavior. (Middle) If the robot behavior does not match the requirements (it hits the red bar in the depicted case), novel demonstrations of the task are provided. (Bottom) The robot executes the refined behavior and avoids the obstacle.

while keeping the rest unchanged. The works in [9,10] propose passivity-based controllers to allow a human operator to incrementally demonstrate a DMP trajectory safely interacting with the robot during the execution. The limitation of these approaches is that they rely on DMPs that are time dependent DS. As quantitatively shown in [15], time dependent DS use task dependent heuristics to preserve the shape of a demonstrated motion in the face of temporal or spacial perturbations. On the other hand, time-independent DS naturally adapt to motion perturbations and exhibit higher generalization capabilities.

In [12], authors propose to reshape the velocity of an autonomous (i.e. time independent) DS using a modulation matrix. The modulation matrix is parametrized as a rotation matrix and a scalar gain. These parameters are incrementally learned from demonstration using Gaussian processes regression [13]. The approach in [12] has the advantage to locally modify the DS dynamics, i.e. the dynamics in regions of the state space far from the demonstrated trajectories remain unchanged. The main limitations of [12] are that it directly applies only to first-order DS and to low-dimensional spaces (up to 3 dimensions where rotations can be uniquely represented). In our previous work [4], we propose to suppress a learned reshaping term (additive control input) using a time-dependent signal. The reshaping term depends on less parameters than the modulation matrix in [12] and it naturally applies to high-dimensional DS and state spaces. Moreover, generated trajectories accurately follow the demonstrated ones. However, the time-dependence introduces practical limitations due to the hand tuning of the time constant used to suppress the reshaping term. For instance, if a longer trajectory is required for some initial conditions, the reshaping term can be suppressed too early introducing deviations from the demonstrated path.

In this paper, we propose a novel parameterization of the reshaping term that preserves the stability of the reshaped DS without introducing time-dependencies. This is obtained by projecting the reshaping term in the subspace orthogonal to the gradient of a given Lyapunov function. The presented parameterization is general and no restrictions are imposed on the order of the DS, as well as on the dimension of the state space. Moreover, the dynamics of the original DS are locally affected by the reshaping action, giving the possibility to learn different behaviors in different regions of the state space. To this end, we adopt a kernel based (local) regression technique, namely Gaussian process regression, to retrieve a smooth control input for each state. The control action is learned incrementally from user demonstrations by deciding when new points are added to the training set. As in [4,12], a trajectory-based sparsity criterion is used to reduce the amount of points added to the training set and reduce the computation time. The incremental learning procedure proposed in this work is shown in Fig. 1.

The rest of the paper is organized as follows. Section 2 presents the theoretical background and the proposed parameterization of the reshaping term. The incremental learning algorithm is described in Sect. 3. Simulations and experiments are presented in Sect. 4. Section 5 states the conclusions and the future extensions.

2 Orthogonal Reshaping of Dynamical Systems

In this section, we describe an approach to reshape the dynamics of a generic DS without modifying its stability properties.

2.1 Theoretical Background

We assume that the robot's task is encoded into a m-th order autonomous DS (time-dependencies are omitted to simplify the notation)

$$\boldsymbol{p}^{(m)} = \boldsymbol{g}\left(\boldsymbol{p}, \boldsymbol{p}^{(1)}, \ldots, \boldsymbol{p}^{(m-1)}\right) \tag{1}$$

where $\boldsymbol{p} \in \mathbb{R}^n$ is the robot position (in joint or Cartesian space), $\boldsymbol{p}^{(i)}$ is the i-th time derivative of $\boldsymbol{p}$ and $\boldsymbol{g} : \mathbb{R}^n \rightarrow \mathbb{R}^n$ is, in general, a non-linear function. The m-th order dynamics (1) can be rewritten, through the change of variables $\boldsymbol{x}^T = [\boldsymbol{x}_1^T, \ldots, \boldsymbol{x}_m^T] = [\boldsymbol{p}^T, \ldots, (\boldsymbol{p}^{(m-1)})^T]$, as the first-order dynamics

$$\begin{cases} \dot{\boldsymbol{x}}_1 = \boldsymbol{x}_2 \\ \quad \cdots \\ \dot{\boldsymbol{x}}_m = \boldsymbol{g}\left(\boldsymbol{x}_1, \ldots, \boldsymbol{x}_m\right) \end{cases} \longrightarrow \dot{\boldsymbol{x}} = \boldsymbol{f}(\boldsymbol{x}) \tag{2}$$

where $\boldsymbol{x} \in \mathbb{R}^{mn}$ is the state vector. The solution of (2) $\boldsymbol{\Phi}(\boldsymbol{x}_0, t) \in \mathbb{R}^{mn}$ is called trajectory. Different initial conditions $\boldsymbol{x}_0$ generate different trajectories.

A point $\hat{\boldsymbol{x}} : \boldsymbol{f}(\hat{\boldsymbol{x}}) = \boldsymbol{0} \in \mathbb{R}^{mn}$ is an equilibrium point. An equilibrium is locally asymptotically stable (LAS) if $\lim_{t \rightarrow +\infty} \boldsymbol{\Phi}(\boldsymbol{x}_0, t) = \hat{\boldsymbol{x}}, \forall \boldsymbol{x}_0 \in S \subset \mathbb{R}^{mn}$. If $S = \mathbb{R}^{mn}$, $\hat{\boldsymbol{x}}$ is globally asymptotically stable (GAS) and it is the only equilibrium of the DS. A sufficient condition for $\hat{\boldsymbol{x}}$ to be GAS is that there exists a scalar, continuously differentiable function of the state variables $V(\boldsymbol{x}) \in \mathbb{R}$ satisfying (3a)–(3c) (see, for example, [11] for further details).

$$V(\boldsymbol{x}) \geq 0,\ \forall \boldsymbol{x} \in \mathbb{R}^{mn} \text{ and } V(\boldsymbol{x}) = 0 \Longleftrightarrow \boldsymbol{x} = \hat{\boldsymbol{x}} \tag{3a}$$

$$\dot{V}(\boldsymbol{x}) \leq 0,\ \forall \boldsymbol{x} \in \mathbb{R}^{mn} \text{ and } \dot{V}(\boldsymbol{x}) = 0 \Longleftrightarrow \boldsymbol{x} = \hat{\boldsymbol{x}} \tag{3b}$$

$$V(\boldsymbol{x}) \to \infty \text{ as } \|\boldsymbol{x}\| \to \infty \text{ (radially unbounded)} \tag{3c}$$

Note that, if condition (3c) is not satisfied, the equilibrium point is LAS. A function satisfying (3a)–(3b) is called a Lyapunov function.

2.2 Reshaping Control Input

If the task consists in reaching a specific position $\hat{\boldsymbol{x}}$ (discrete movement), one can assume that (2) has a GAS equilibrium at $\hat{\boldsymbol{x}}$ and that a Lyapunov function $V(\boldsymbol{x})$ is known [6,14]. Let us consider the reshaped DS in the form

$$\dot{\boldsymbol{x}} = \boldsymbol{f}(\boldsymbol{x}) + \boldsymbol{u}(\boldsymbol{x}) \tag{4}$$

where $\boldsymbol{u}(\boldsymbol{x}) = [\boldsymbol{0}, \ldots, \boldsymbol{0}, \boldsymbol{u}_m(\boldsymbol{x})] \in \mathbb{R}^{mn}$ is a continuous control input that satisfies

$$\boldsymbol{u}(\hat{\boldsymbol{x}}) = \boldsymbol{0} \Longleftrightarrow \boldsymbol{u}_m(\hat{\boldsymbol{x}}) = \boldsymbol{0} \in \mathbb{R}^n \tag{5a}$$

$$V_x \boldsymbol{u}(\boldsymbol{x}) = 0 \Longleftrightarrow V_{x_m} \boldsymbol{u}_m(\boldsymbol{x}) = \frac{\partial V}{\partial \boldsymbol{x}_m} \boldsymbol{u}_m(\boldsymbol{x}) = 0, \forall \boldsymbol{x}_m \in \mathbb{R}^n \tag{5b}$$

where V_x indicates the gradient of $V(\boldsymbol{x})$ with respect to $\boldsymbol{x}$, i.e. $V_x = \partial V(\boldsymbol{x})/\partial \boldsymbol{x}$.

Under conditions (5) the following theorem holds:

Theorem 1. *A GAS equilibrium $\hat{\boldsymbol{x}}$ of (2) is also a GAS equilibrium of the reshaped DS (4).*

Proof. From (4) and (5a) it holds that $\boldsymbol{f}(\hat{\boldsymbol{x}}) + \boldsymbol{u}(\hat{\boldsymbol{x}}) = \boldsymbol{0}$, i.e. $\hat{\boldsymbol{x}}$ is an equilibrium of (4). To analyze the stability of $\hat{\boldsymbol{x}}$, let us consider $V(\boldsymbol{x})$, the Lyapunov function for (2), as a candidate Lyapunov function for (4). Being $V(\boldsymbol{x})$ a Lyapunov function for (2) it satisfies conditions (3a) and (3c) also for the reshaped DS (4). The condition (3b) can expressed in terms of the gradient of $V(\boldsymbol{x})$ as

$$\begin{aligned}\dot{V}(\boldsymbol{x}) = V_x \dot{\boldsymbol{x}} &= \left[\frac{\partial V(\boldsymbol{x})}{\partial \boldsymbol{x}_1}, \ldots, \frac{\partial V(\boldsymbol{x})}{\partial \boldsymbol{x}_m}\right] (\boldsymbol{f}(\boldsymbol{x}) + \boldsymbol{u}(\boldsymbol{x})) \\ &= \frac{\partial V(\boldsymbol{x})}{\partial \boldsymbol{x}_1} \boldsymbol{x}_2 + \ldots + \frac{\partial V(\boldsymbol{x})}{\partial \boldsymbol{x}_m} \boldsymbol{g}(\boldsymbol{x}) + \cancel{\frac{\partial V(\boldsymbol{x})}{\partial \boldsymbol{x}_m} \boldsymbol{u}_m(\boldsymbol{x})} = V_x \boldsymbol{f}(\boldsymbol{x}) < 0, \forall \boldsymbol{x} \neq \hat{\boldsymbol{x}}\end{aligned}$$

where $(\partial V(\boldsymbol{x})/\partial \boldsymbol{x}_m) \boldsymbol{u}_m(\boldsymbol{x}) = 0$ by assumption (5b). □

Corollary 1. *Theorem 1 implies that $\hat{\boldsymbol{x}}$ is the only equilibrium point of (4), i.e. $\boldsymbol{f}(\boldsymbol{x}) + \boldsymbol{u}(\boldsymbol{x})$ vanishes only at $\boldsymbol{x} = \hat{\boldsymbol{x}}$.*

Proof. In Theorem 1 it is proved that $V_x(\boldsymbol{f}(\boldsymbol{x}) + \boldsymbol{u}(\boldsymbol{x}))$ vanishes only at $\hat{\boldsymbol{x}}$. Hence, also $\boldsymbol{f}(\boldsymbol{x}) + \boldsymbol{u}(\boldsymbol{x})$ vanishes only at $\hat{\boldsymbol{x}}$. □

Corollary 2. *Theorem 1 still holds for a LAS equilibrium, i.e. if $\hat{\boldsymbol{x}}$ is a LAS equilibrium of (2) then $\hat{\boldsymbol{x}}$ is a LAS equilibrium of (4).*

Proof. If $\hat{\boldsymbol{x}}$ is LAS, only the conditions (3a)–(3b) are satisfied $\forall \boldsymbol{x} \in S \subset \mathbb{R}^{mn}$. The proof of Theorem (1) still holds if $\boldsymbol{x} \in S \subset \mathbb{R}^{mn}$. □

Theorem 1 has a clear physical interpretation for second-order dynamical systems in the form $\dot{\boldsymbol{x}}_1 = \boldsymbol{x}_2$, $\dot{\boldsymbol{x}}_2 = \boldsymbol{K}(\hat{\boldsymbol{x}}_1 - \boldsymbol{x}_1) - \boldsymbol{D}\boldsymbol{x}_2$, where $\boldsymbol{K}$ and $\boldsymbol{D}$ are positive definite matrices and $\hat{\boldsymbol{x}} = [\hat{\boldsymbol{x}}_1^T, \boldsymbol{0}^T]^T$ is the equilibrium point. The GAS of $\hat{\boldsymbol{x}}$ can be proven through the energy-based Lyapunov function $V = \frac{1}{2}(\hat{\boldsymbol{x}}_1 - \boldsymbol{x}_1)^T\boldsymbol{K}(\hat{\boldsymbol{x}}_1 - \boldsymbol{x}_1) + \frac{1}{2}\boldsymbol{x}_2^T\boldsymbol{x}_2$ and the La Salle's theorem [11]. The assumptions on $\boldsymbol{u}_m$ in Theorem 1 can be satisfied by choosing $\boldsymbol{u}_m$ orthogonal to $V_{x_2} = \boldsymbol{x}_2$. Hence, $\boldsymbol{u}_m$ is a force that does no work (orthogonal to the velocity), i.e. $\boldsymbol{u}_m$ modifies the trajectory of the system but not its energy.

2.3 Control Input Parametrization

In order to satisfy the conditions (5) we choose the control input $\boldsymbol{u}_m$ in (4) as

$$\boldsymbol{u}_m = \begin{cases} \boldsymbol{N}\boldsymbol{u}_d = \boldsymbol{N}h(\boldsymbol{x}_1)\left(\boldsymbol{p}_d(\boldsymbol{x}_1) - \boldsymbol{x}_1\right) & \text{if } \boldsymbol{x} \neq \hat{\boldsymbol{x}} \\ \boldsymbol{0} & \text{otherwise} \end{cases} \tag{6}$$

where $\boldsymbol{x}_1 \in \mathbb{R}^n$ is the position of the robot. The scalar gain $h(\boldsymbol{x}_1) \geq 0$ and the desired position $\boldsymbol{p}_d(\boldsymbol{x}_1) \in \mathbb{R}^n$ are learned from demonstrations (see Sect. 3). The adopted parametrization (6) requires always $n+1$ parameters, i.e. the position vector $\boldsymbol{p}_d \in \mathbb{R}^n$ (where n is the Cartesian or joint space dimension) and the scalar gain $h \in \mathbb{R}$. For comparison, consider that the parametrization in [12] uses a rotation and a scalar gain and that a minimal representation of the orientation in $\mathbb{R}^n$ requires at least $n(n-1)/2$ parameters [25]. The vector $\boldsymbol{u}_d$ represents an elastic force attracting the position $\boldsymbol{x}_1$ towards the desired position $\boldsymbol{p}_d$. The matrix $\boldsymbol{N}$ is used to project $\boldsymbol{u}_d$ into the subspace orthogonal to V_{x_m} and it is defined as

$$\boldsymbol{N} = \|V_{x_m}\|^2 \left(\boldsymbol{I}_{n\times n} - \bar{V}_{x_m}^T \bar{V}_{x_m}\right) \tag{7}$$

where $\boldsymbol{I}_{n\times n}$ is the n-dimensional identity matrix and $\bar{V}_{x_m} = V_{x_m}/\|V_{x_m}\|$. The term $\|V_{x_m}\|^2$ guarantees a smooth convergence of $\boldsymbol{u}_m$ to zero. Note that the control input $\boldsymbol{u}_m \to \boldsymbol{0}$ if $h(\boldsymbol{x}_1) \to 0$. This property is exploited in Sect. 3 to locally modify the trajectory of the DS.

3 Learning Reshaping Terms

In this section, an approach is described to learn and online retrieve for each position $\boldsymbol{x}_1$ the parameter vector $\boldsymbol{\lambda} = [h,\ \boldsymbol{p}_d^T]$ that parametrizes the control input in (6). We use a local regression technique, namely Gaussian process regression (GPR), to ensure that $\boldsymbol{u}_m \to \boldsymbol{0}$ ($h \to 0$) when the robot is far from the demonstrated trajectories. This makes it possible to locally follow the demonstrated trajectories, leaving the rest almost unchanged.

3.1 Compute Training Data

Consider that a new demonstration of a task is given as $\boldsymbol{X} = \{\boldsymbol{x}_{d,1}^t, \dot{\boldsymbol{x}}_{d,m}^t\}_{t=1}^T$, where $\boldsymbol{x}_{d,1}^t \in \mathbb{R}^n$ is the desired position at time t and $\dot{\boldsymbol{x}}_{d,m}^t \in \mathbb{R}^n$ is the time derivative of the last state component $\boldsymbol{x}_{d,m}^t$ at time t. For example, if one wants to reshape a second-order DS, then $\boldsymbol{X}$ contains T positions and T accelerations. The procedure to transform the demonstration into T observations of $\boldsymbol{\Lambda} = \{\boldsymbol{\lambda}^t = [\lambda_1, \ldots, \lambda_{n+1}]^t = [h, \, \boldsymbol{p}_d^T]^t\}_{t=1}^T$ requires following steps:

I. Set $\{\boldsymbol{p}_d^t = \boldsymbol{x}_{d,1}^t\}_{t=1}^T$, where $\boldsymbol{x}_{d,1}$ are the demonstrated positions.

The gain h^t in (6) multiplies the position error $\boldsymbol{p}_d - \boldsymbol{x}_1$ and it is used to modulate the control action $\boldsymbol{u}_m$ and to improve the overall tracking performance. The value of h^t is computed by considering that $\dot{\boldsymbol{x}}_m = \boldsymbol{g}(\boldsymbol{x}) + \boldsymbol{u}_m$ from (2) and (4). The following steps are needed:

II. Create the initial state vector $\boldsymbol{x}_d^0 = [(\boldsymbol{x}_{d,1}^1)^T, \boldsymbol{0}^T, \ldots, \boldsymbol{0}^T]^T \in \mathbb{R}^{mn}$.
III. Compute $\left\{\boldsymbol{x}_o^t = \boldsymbol{\Phi}(\boldsymbol{x}_d^0, (t-1)\delta t)\right\}_{t=1}^T$, where $\boldsymbol{\Phi}$ is the solution of (2) with initial condition $\boldsymbol{x}_d^0$ (see Sect. 2.1) and δt is the sampling time.
IV. Compute $\{\boldsymbol{u}_m^t\}_{t=1}^T$ from (6) with $h = 1$ and $\{\boldsymbol{x}_1^t = \boldsymbol{x}_{o,1}^t\}_{t=1}^T$.
V. Set

$$h^t = \begin{cases} \dfrac{\|\dot{\boldsymbol{x}}_{d,m}^t - \boldsymbol{g}(\boldsymbol{x}_o^t)\|}{\|\boldsymbol{u}_m^t\|} & \text{if } \|\boldsymbol{u}_m^t\| > 0 \\ 0 & \text{otherwise} \end{cases} \quad t = 1, \ldots, T \tag{8}$$

Once the observations of $\boldsymbol{\lambda}$ are computed, any local regression technique can be applied to learn the relationship between $\boldsymbol{\lambda}$ and the position $\boldsymbol{x}_1$ of the DS to reshape.

3.2 Gaussian Process Regression

Gaussian processes (GP) are widely used to learn input-output mappings from observations [13]. GP models the scalar noisy process $\lambda^t = f(\boldsymbol{x}_1^t) + \epsilon \in \mathbb{R}, t = 1, \ldots, T$ with ϵ a Gaussian noise with zero mean and variance σ_n^2. Therefore, n processes λ_i^t are assumed to generate the training input $\boldsymbol{X} = \{\boldsymbol{x}_1^t\}_{t=1}^T$ and output $\boldsymbol{\Lambda}_i = \{\lambda_i^t\}_{t=1}^T$. Given the training pairs $(\boldsymbol{X}, \boldsymbol{\Lambda}_i)$ and a query point $\boldsymbol{x}^*$, it is possible to compute the joint distribution

$$\begin{bmatrix} \boldsymbol{\Lambda}_i \\ \lambda_i^* \end{bmatrix} \sim \mathcal{N}\left(\boldsymbol{0}, \begin{bmatrix} \boldsymbol{K}_{XX} + \sigma_n^2 \boldsymbol{I} & \boldsymbol{K}_{x^*X} \\ \boldsymbol{K}_{Xx^*} & k(\boldsymbol{x}^*, \boldsymbol{x}^*) \end{bmatrix}\right) \tag{9}$$

where λ_i^* is the expected output at $\boldsymbol{x}^*$. The matrix $\boldsymbol{K}_{x^*X} = \{k(\boldsymbol{x}^*, \boldsymbol{x}_1^t)\}_{t=1}^T$, $\boldsymbol{K}_{Xx^*} = \boldsymbol{K}_{x^*X}^T$. Each element ij of $\boldsymbol{K}_{XX}$ is given by $\{\boldsymbol{K}_{XX}\}_{ij} = k(\boldsymbol{x}_i, \boldsymbol{x}_j)$, where $k(\bullet, \bullet)$ is a user-defined covariance function. In this work, we used the squared exponential covariance function

$$k(\boldsymbol{x}_i, \boldsymbol{x}_j) = \sigma_k^2 \exp\left(-\frac{\|\boldsymbol{x}_i - \boldsymbol{x}_j\|^2}{2l}\right) + \sigma_n^2 \delta(\boldsymbol{x}_i, \boldsymbol{x}_j) \tag{10}$$

$k(\boldsymbol{x}_i, \boldsymbol{x}_j)$ is parameterized by the 3 positive parameters σ_k^2, σ_n^2, and l. The tunable parameters σ_k^2, σ_n^2, and l can be hand-crafted or learned from training data [13]. We decide to keep them fixed in order to perform incremental learning by simply adding new points to the training set. It is worth noticing that the adopted kernel function (10) guarantees that $\boldsymbol{\lambda} \to \boldsymbol{0}$ for points far from the demonstrated positions.

Predictions with a GP model are made using the conditional distribution of $\lambda_i^*|\boldsymbol{\Lambda}_i$, i.e.

$$\lambda_i^*|\boldsymbol{\Lambda}_i \sim \mathcal{N}\left(\mu_{\lambda_i^*|\boldsymbol{\Lambda}_i}, \sigma^2_{\lambda_i^*|\boldsymbol{\Lambda}_i}\right) \tag{11}$$

where

$$\begin{aligned} \mu_{\lambda_i^*|\boldsymbol{\Lambda}_i} &= \boldsymbol{K}_{x^*X}\left(\boldsymbol{K}_{XX} + \sigma_n^2\boldsymbol{I}\right)^{-1}\boldsymbol{\Lambda}_i \\ \sigma^2_{\lambda_i^*|\boldsymbol{\Lambda}_i} &= k(\boldsymbol{x}^*, \boldsymbol{x}^*) - \boldsymbol{K}_{x^*X}\left(\boldsymbol{K}_{XX} + \sigma_n^2\boldsymbol{I}\right)^{-1}\boldsymbol{K}_{Xx^*} \end{aligned} \tag{12}$$

The mean $\mu_{\lambda_i^*|\boldsymbol{\Lambda}_i}$ approximates λ_i^*, while the variance $\sigma^2_{\lambda_i^*|\boldsymbol{\Lambda}_i}$ plays the role of a confidence bound. If, as in this work, a multidimensional output is considered, one can simply train one GP for each dimension.

To reduce the computation effort due to the matrix inversion in (12), incremental GP algorithms introduce criteria to sparsely represent incoming data [22]. Assuming that T data $\{\boldsymbol{x}_1^t,\ h^t,\ \boldsymbol{p}_d^t\}_{t=1}^T$ are already in the training set, we add a new data point $[\boldsymbol{x}_1^{T+1},\ h^{T+1},\ \boldsymbol{p}_d^{T+1}]$ if the cost

$$C^{T+1} = \|\boldsymbol{p}_d^{T+1} - \hat{\boldsymbol{p}}_d^{T+1}\| > \bar{c} \tag{13}$$

where $\hat{\boldsymbol{p}}_d^{T+1}$ indicates the position predicted at $\boldsymbol{x}_1^{T+1}$ with (12) using only data $\{\boldsymbol{x}_1^t,\ h^t,\ \boldsymbol{p}_d^t\}_{t=1}^T$ already in the training set. Similarly to [4,12], the tunable parameter $\bar{c}$ represents the error in approximating demonstrated positions and it can be easily tuned. For example, $\bar{c} = 0.2$ means that position errors smaller than 0.2 m are acceptable. The proposed incremental reshaping approach is summarized in Table 1.

4 Results

4.1 Simulation - Learning Bi-Modal Behaviors

The goal of this simulation is to illustrate the incremental nature of the proposed reshaping approach, its ability to learn different behaviors in different regions of the space, and the possibility to reshape high order DS. The original trajectory is obtained by numerically integrating ($\delta t = 0.01$ s) the second-order DS

$$\begin{aligned} \dot{\boldsymbol{x}}_1 &= \boldsymbol{x}_2 \\ \dot{\boldsymbol{x}}_2 &= -10\boldsymbol{x}_1 - 2\sqrt{10}\boldsymbol{x}_2 \end{aligned} \tag{14}$$

where $\boldsymbol{x}_1 = [x, y]^T \in \mathbb{R}^2$ is the position and $\boldsymbol{x}_2$ the velocity. The system (14) has a GAS equilibrium at $\hat{\boldsymbol{x}} = \boldsymbol{0} \in \mathbb{R}^4$ and Lyapunov function $V(\boldsymbol{x}) = \frac{1}{2}(\boldsymbol{x}_1^T\boldsymbol{x}_1 + \boldsymbol{x}_2^T\boldsymbol{x}_2)$.

Table 1. Proposed reshaping approach.

Batch	Create a set of predefined tasks encoded as stable DS Stable DS can be designed by the user or learned from demonstrations as in [6,14]
	Provide a Lyapunov function $V(\boldsymbol{x})$ for each DS
Incremental	Observe the robot's behavior in novel scenarios
	If needed, provide a corrective demonstration, for example by kinesthetic teaching the robot
	Learn the parameters of the control input (4), as described in Sect. 3. Tuning parameters can be set empirically by simulating the reshaped DS
	Repeat until the refined behavior is satisfactory

Local demonstrations are drawn from different Gaussian distributions, as described in Table 2, to obtain different bi-modal behaviors. A total of four demonstrations are used in each case, i.e. two (red and green crosses in Fig. 2) for the behavior in the region $\mathcal{R}^+$ where $x > 0$, two (magenta and blue crosses in Fig. 2) for the region $\mathcal{R}^-$ where $x < 0$. As shown in Fig. 2, the original DS position trajectory (black solid line) is incrementally adapted to follow the demonstrated positions and different behaviors are effectively learned in $\mathcal{R}^+$ and $\mathcal{R}^-$. Totally four simulations are conducted and shown in Fig. 2. In all the presented cases, the DS is successfully reshaped to follow the demonstrated trajectories. The proposed approach locally modifies the DS, in fact demonstrations in $\mathcal{R}^+$, being far from $\mathcal{R}^-$, do not affect the behavior in $\mathcal{R}^-$ (and vice versa). The equilibrium position $\hat{\boldsymbol{x}}_1 = \boldsymbol{0} \in \mathbb{R}^2$ is always reached, as expected from Theorem 1. Results are obtained with noise variance $\sigma_n^2 = 0.1$, signal variance $\sigma_k^2 = 1$, length scale $l = 0.4$, and threshold $\bar{c} = 0.02$ m.

4.2 Experiments

The effectiveness of the proposed approach is demonstrated with two experiments on a KUKA LWR IV+ 7 DoF manipulator. In both experiments, novel demonstrations of the desired position are provided to the robot by kinesthetic teaching. To guarantee a safe physical guidance, the task is interrupted and the robot is put in the gravity compensation mode as soon as the user touches the robot. The external torques estimation provided by the fast research interface [23] is used to detect physical contacts.

End-Effector Collision Avoidance. This experiment shows the ability of the proposed reshaping approach to learn different behaviors in different regions of the space and the possibility to reshape non-linear DS. The task consists in reaching the goal position $\hat{\boldsymbol{x}}_1 = [-0.52, 0, 0.02]^T$ m with the robot's end-effector

Table 2. Demonstrations used for the four simulations in Fig. 2.

Figure	Demonstrations in $\mathcal{R}^-$	Demonstrations in $\mathcal{R}^+$
2(a)	$x \in [-2, -2.5]$ $y = 0.1\mathcal{N}(-2, 0.03)$	$x \in [2, 2.5]$ $y = 0.1\mathcal{N}(2, 0.03)$
2(b)	$x \in [-2, -2.5]$ $y = -0.1\mathcal{N}(-2, 0.03)$	$x \in [2, 2.5]$ $y = 0.1\mathcal{N}(2, 0.03)$
2(c)	$x \in [-2, -2.5]$ $y = 0.1\mathcal{N}(-2, 0.03)$	$x \in [2, 2.5]$ $y = -0.1\mathcal{N}(2, 0.03)$
2(d)	$x \in [-2, -2.5]$ $y = -0.1\mathcal{N}(-2, 0.03)$	$x \in [2, 2.5]$ $y = -0.1\mathcal{N}(2, 0.03)$

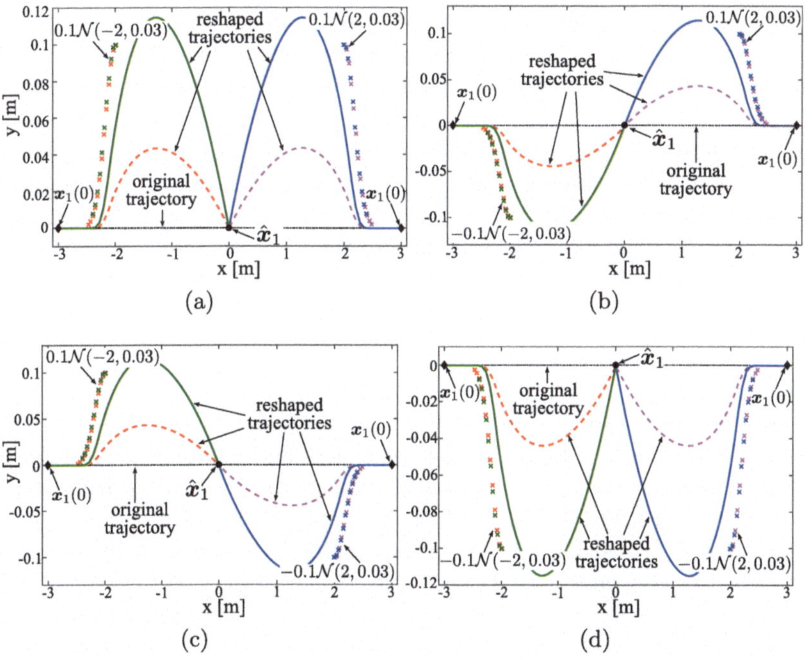

Fig. 2. Different bi-modal behaviors obtained by reshaping the same dynamical system. Red and magenta dashed lines are the reshaped trajectories after providing two demonstrations, one (red crosses) for $\mathcal{R}^+$ and one (magenta crosses) for $\mathcal{R}^-$. Blue and green solid lines are the reshaped trajectories after providing four demonstrations.

while avoiding a box (see Fig. 3(d)) of size $7 \times 7 \times 23$ cm. Two boxes are placed in the scene in different positions, one in the region $\mathcal{R}^+$ where $y > 0$ (Fig. 3(d)), one in the region $\mathcal{R}^-$ where $y < 0$ (Fig. 3(e)). Hence, the robot has to learn a bi-modal behavior to avoid collisions in $\mathcal{R}^+$ and $\mathcal{R}^-$.

The original position trajectory is obtained by numerically integrating ($\delta t = 0.005\,$s) the first-order and non-linear DS $\dot{\boldsymbol{x}}_1 = \boldsymbol{f}(\boldsymbol{x}_1 - \hat{\boldsymbol{x}}_1)$, where $\boldsymbol{x}_1 = [x, y, z]^T \in \mathbb{R}^3$ is the end-effector position and $\dot{\boldsymbol{x}}_1 \in \mathbb{R}^3$ is the end-effector linear velocity. The orientation is kept fixed. The original DS is learned from demonstrations by using the approach in [14]. The Lyapunov function for the original DS is $V = \frac{1}{2}(\boldsymbol{x}_1 - \hat{\boldsymbol{x}}_1)^T(\boldsymbol{x}_1 - \hat{\boldsymbol{x}}_1)$ [14]. The original end-effector trajectories are shown in Fig. 3(a)–(c) (black solid lines).

Following the original trajectory generated with initial position $\boldsymbol{x}_1(0) = [-0.52, 0.5, 0.02]^T$ m (or $\boldsymbol{x}_1(0) = [-0.52, -0.5, 0.02]^T$ m), the robot hits the box. To prevent this, two partial demonstrations (one in $\mathcal{R}^+$ and one $\mathcal{R}^-$) are provided to show to the robot how to avoid the obstacles (brown solid lines in Fig. 3(a)–(c)). The original DS position trajectories (black solid lines in Fig. 3(a)–(c)) are incrementally adapted to follow the demonstrated positions and different avoiding behaviors are effectively learned in $\mathcal{R}^+$ and $\mathcal{R}^-$.

The proposed approach locally modifies the DS, in fact demonstrations in $\mathcal{R}^+$, being far from $\mathcal{R}^-$, do not affect the behavior in $\mathcal{R}^-$ (and vice versa). The equilibrium position $\hat{\boldsymbol{x}}_1 = [-0.52, 0, 0.02]^T$ m is always reached, as stated by Theorem 1. Figure 3 also shows the learned behaviors (green in $\mathcal{R}^+$ and blue in $\mathcal{R}^-$ solid lines) for different initial positions in a 3D view (Fig. 3(a)) and in the xz plane (Fig. 3(b) and (c)). In all cases, the robot is able to achieve the task. Snapshots of the learned bi-modal behavior are depicted in Fig. 3(d) and (e). Results are obtained with noise variance $\sigma_n^2 = 0.1$, signal variance $\sigma_k^2 = 1$, length scale $l = 0.001$, and the threshold $\bar{c} = 0.04\,$m. With the adopted $\bar{c}$ only 106 points over 798 are added to the GP.

Joint Space Collision Avoidance. This experiment shows the scalability of the proposed approach to high dimensional spaces and its ability to reshape high order DS. The task is a point-to-point motion in the joint space from $\boldsymbol{x}_1(0) = [35, 55, 15, -65, -15, 50, 90]^T$ deg to $\hat{\boldsymbol{x}}_1 = [-60, 30, 30, -70, -15, 85, 15]^T$ deg. The original joint position trajectory is obtained by numerically integrating ($\delta t = 0.005\,$s) the second-order DS $\dot{\boldsymbol{x}}_1 = \boldsymbol{x}_2$, $\dot{\boldsymbol{x}}_2 = 2(\hat{\boldsymbol{x}}_1 - \boldsymbol{x}_1) - 2\sqrt{2}\boldsymbol{x}_2$, where $\boldsymbol{x}_1 = [\theta_1, \ldots, \theta_7]^T \in \mathbb{R}^7$ are the joint angles and $\boldsymbol{x}_2 \in \mathbb{R}^7$ the joint velocities. The system has a GAS equilibrium at $\hat{\boldsymbol{x}} = [\hat{\boldsymbol{x}}_1^T, \boldsymbol{0}^T]^T \in \mathbb{R}^{14}$ and Lyapunov function $V = \frac{1}{2}(\hat{\boldsymbol{x}}_1 - \boldsymbol{x}_1)^T(\hat{\boldsymbol{x}}_1 - \boldsymbol{x}_1) + \frac{1}{2}\boldsymbol{x}_2^T\boldsymbol{x}_2$. The original joint angle trajectories are shown in Fig. 4 (black solid lines).

As shown in Fig. 1, following the original trajectory the robot hits an unforeseen obstacle (the red bar in Fig. 1). A kinesthetic demonstration is then provided (red lines in Fig. 4) to avoid the collision. With the reshaped trajectory (blue lines in Fig. 4) the robot is able to avoid the obstacle (Fig. 1) and to reach the desired goal $\hat{\boldsymbol{x}}_1$. Results are obtained with noise variance $\sigma_n^2 = 0.1$, signal variance $\sigma_k^2 = 1$, length scale $l = 0.01$, and the threshold $\bar{c} = 15\,$deg. With the adopted $\bar{c}$ only 27 points over 178 are added to the GP.

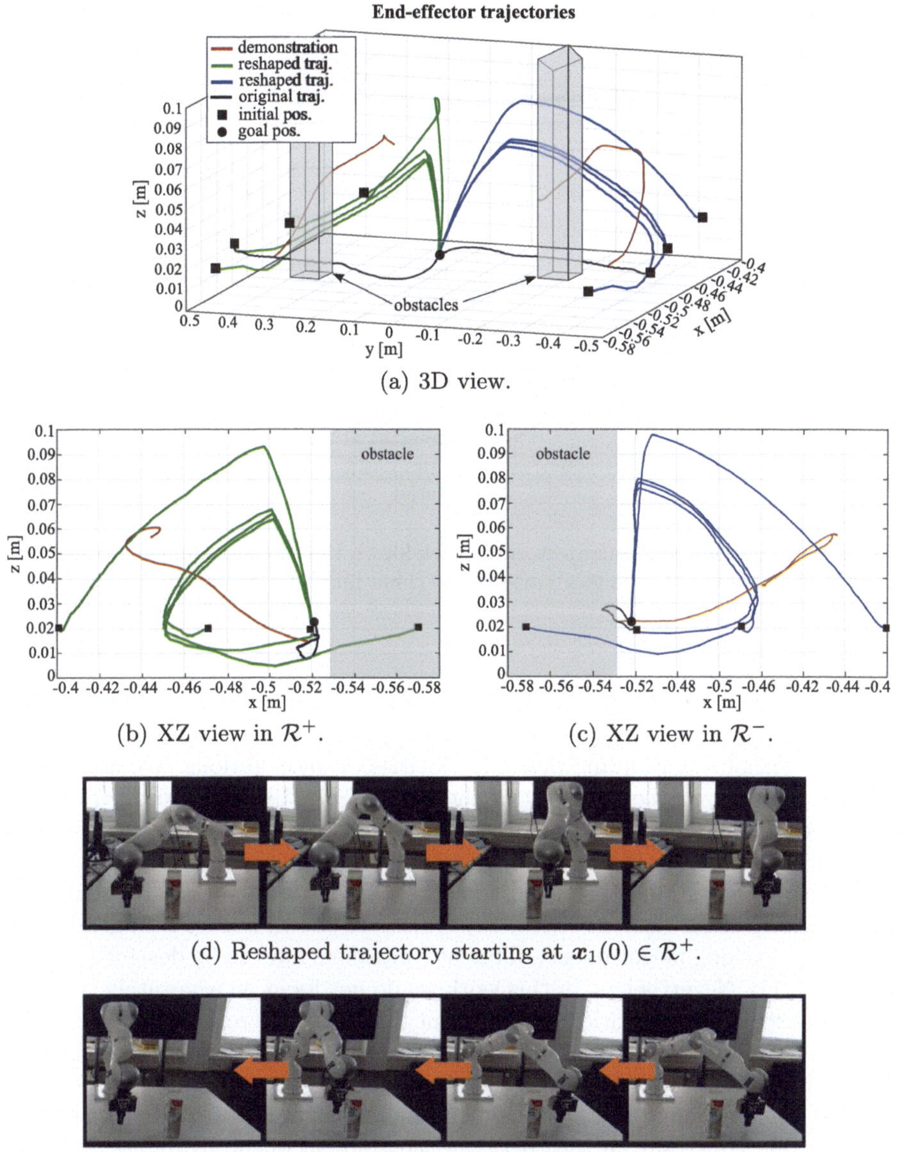

(a) 3D view.

(b) XZ view in $\mathcal{R}^+$.

(c) XZ view in $\mathcal{R}^-$.

(d) Reshaped trajectory starting at $\boldsymbol{x}_1(0) \in \mathcal{R}^+$.

(e) Reshaped trajectory starting at $\boldsymbol{x}_1(0) \in \mathcal{R}^-$.

Fig. 3. Results of the end-effector collision avoidance experiment.

4.3 Discussion

The proposed approach works for high order DS, as underlined in Sect. 2 and demonstrated in Sect. 4.2. Being robot manipulators dynamics described by second-order DS, a second-order DS is sufficient to generate dynamically feasible

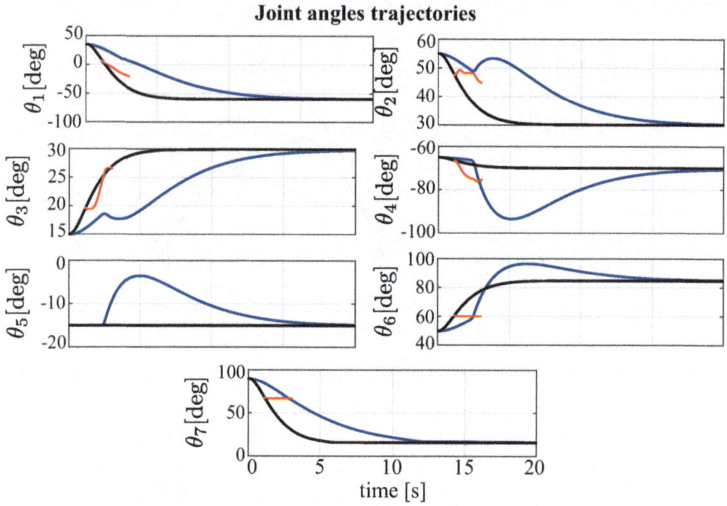

Fig. 4. Original joint angles trajectories (black lines), the provided demonstration (red lines) and reshaped joint angles trajectories (blue lines) for the joint space collision avoidance experiment.

trajectories. For this reason, we show results for DS up to the second order. The adopted control law in (6) pushes the robot position towards the demonstrated position, without considering desired velocities or accelerations. We adopt this solution because, in the majority of the cases, a user is interested in reconfiguring the robot and (s)he can hardly show a desired velocity (or acceleration) behavior through kinesthetic teaching.

It must be noted that the proposed control law (6) does not always guarantee good tracking of the demonstrated trajectories, as shown in Fig. 4. In general, to have good tracking performance, different controllers have to be designed for different DS [11]. Nevertheless, in this work we do not focus on accurately tracking the demonstrated trajectories, but we want to modify the robot's behavior until the task is correctly executed. The joint angle trajectories in Fig. 4 guarantee the correct execution of the task, i.e. the robot converges to the desired joint position while avoiding the obstacle. The loss of accuracy also depends on the orthogonality constraint in (5b) between the gradient of the Lyapunov function and the control input. This constrain allows only motions perpendicular to the gradient to be executed, which limits the control capabilities and increases the number of demonstrations needed in order to obtain the satisfactory behavior. In principle, it is possible to relax the constraint (5b) by requiring that $V_x \boldsymbol{u}(\boldsymbol{x}) \geq 0$. The design of a control input the satisfies $V_x \boldsymbol{u}(\boldsymbol{x}) \leq 0$ is left as future work.

Figure 4 shows an overshoots in the resulting position trajectory (see, for instance, the angle θ_6). To better understand this behavior, consider that we are controlling a spring-damper (linear) DS with a proportional controller (with a non-linear gain). In case the resulting closed-loop system is not critically damped, the retrieved trajectory overshoots the goal position. For the experiment in Sect. 4.2, adding a damping control action (PD-like controller) would solve the overshoot problem. However, there is no guarantee that a generic non-linear DS does not overshoot under a PD-like control action. Moreover, adding another term to (6) will increase the number of parameters to learn. Therefore, we use a proportional controller in this work and leave the learning of more sophisticated controllers as a future extension.

5 Conclusions and Future Work

We presented a novel approach to incrementally modify the position trajectory of a generic dynamical system, useful to on-line adapt predefined tasks to different scenarios. Compared to state-of-the-art approaches, our method works also for high-order dynamical systems, preserves the time-independence of the DS, and does not affect the stability properties of the reshaped dynamical system, as shown in the conducted Lyapunov-based stability analysis.

A control law is proposed that locally modifies the trajectory of the dynamical system to follow a desired position. Desired positions, as well as the control gain, are learned from demonstrations and retrieved on-line using Gaussian process regression. The procedure is incremental, meaning that the user can add novel demonstrations until the learned behavior is not satisfactory. Due to the local nature of the reshaping control input, different behaviors can be learned and executed in different regions of the space. Simulations and experiments show the effectiveness of the proposed approach in reshaping non-linear, high-order dynamical systems, and its scalability to high dimensional spaces (up to $\mathbb{R}^{14}$).

Our approach applies to dynamical systems with a LAS or a GAS equilibrium point. Nevertheless, DS that converges towards periodic orbits (limit cycles) have been used in robotic applications to generate periodic behaviors [6]. Compared to static equilibria, limit cycles stability has a different characterizations in terms of Lyapunov analysis. Our next research will focus on considering incremental reshaping of periodic motions while preserving their stability properties.

References

1. Saveriano, M., Lee, D.: Point cloud based dynamical system modulation for reactive avoidance of convex and concave obstacles. In: International Conference on Intelligent Robots and Systems, pp. 5380–5387 (2013)
2. Saveriano, M., Lee, D.: Distance based dynamical system modulation for reactive avoidance of moving obstacles. In: International Conference on Robotics and Automation, pp. 5618–5623 (2014)
3. Blocher, C., Saveriano, M., Lee, D.: Learning stable dynamical systems using contraction theory. In: Ubiquitous Robots and Ambient Intelligence, pp. 124–129 (2017)
4. Saveriano, M., Lee, D.: Incremental skill learning of stable dynamical systems. In: International Conference on Intelligent Robots and Systems, pp. 6574–6581 (2018)

5. Saveriano, M., Hirt, F., Lee, D.: Human-aware motion reshaping using dynamical systems. Pattern Recogn. Lett. **99**, 96–104 (2017)
6. Ijspeert, A., Nakanishi, J., Pastor, P., Hoffmann, H., Schaal, S.: Dynamical movement primitives: learning attractor models for motor behaviors. Neural Comput. **25**(2), 328–373 (2013)
7. Khansari-Zadeh, S.M., Billard, A.: A dynamical system approach to realtime obstacle avoidance. Auton. Rob. **32**(4), 433–454 (2012)
8. Karlsson, M., Robertsson, A., Johansson, R.: Autonomous interpretation of demonstrations for modification of dynamical movement primitives. In: International Conference on Robotics and Automation, pp. 316–321 (2017)
9. Talignani Landi, C., Ferraguti, F., Fantuzzi, C., Secchi, C.: A passivity-based strategy for coaching in human–robot interaction. In: International Conference on Robotics and Automation, pp. 3279–3284 (2018)
10. Kastritsi, T., Dimeas, F., Doulgeri, Z.: Progressive automation with DMP synchronization and variable stiffness control. Robot. Autom. Lett. **3**(4), 3279–3284 (2018)
11. Slotine, J.J.E., Li, W.: Applied Nonlinear Control. Prentice-Hall, Upper Saddle River (1991)
12. Kronander, K., Khansari-Zadeh, S.M., Billard, A.: Incremental motion learning with locally modulated dynamical systems. Robot. Auton. Syst. **70**, 52–62 (2015)
13. Rasmussen, C.E., Williams, C.K.I.: Incremental Gaussian Processes for Machine Learning. MIT Press, Cambridge (2006)
14. Khansari-Zadeh, S.M., Billard, A.: Learning stable non-linear dynamical systems with gaussian mixture models. Trans. Robot. **27**(5), 943–957 (2011)
15. Gribovskaya, E., Khansari-Zadeh, S.M., Billard, A.: Learning non-linear multivariate dynamics of motion in robotic manipulators. Int. J. Robot. Res. **30**(1), 80–117 (2011)
16. Saveriano, M., An, S., Lee, D.: Incremental kinesthetic teaching of end-effector and null-space motion primitives. In: International Conference on Robotics and Automation, pp. 3570–3575 (2015)
17. Saveriano, M., Franzel, F., Lee, D.: Merging position and orientation motion primitives. In: International Conference on Robotics and Automation, pp. 7041–7047 (2019)
18. Saveriano, M., Lee, D.: Learning motion and impedance behaviors from human demonstrations. In: International Conference on Ubiquitous Robots and Ambient Intelligence, pp. 368–373 (2014)
19. Lee, D., Ott, C.: Incremental kinesthetic teaching of motion primitives using the motion refinement tube. Autonom. Rob. **31**(2), 115–131 (2011)
20. Billard, A., Calinon, S., Dillmann, R., Schaal, S.: Robot Programming by Demonstration. Springer Handbook of Robotics, pp. 1371–1394 (2008)
21. Calinon, S., Guenter, F., Billard, A.: On learning, representing, and generalizing a task in a humanoid robot. Trans. Syst. Man Cybern. Part B: Cybern. **37**(2), 286–298 (2007)
22. Csató, L.: Gaussian processes - iterative sparse approximations. Ph.D. dissertation, Aston University (2002)
23. Schreiber, G., Stemmer, A., Bischoff, R.: The fast research interface for the KUKA lightweight robot. In: ICRA Workshop on Innovative Robot Control Architectures for Demanding (Research) Applications - How to Modify and Enhance Commercial Controllers, pp. 15–21 (2010)

24. Calinon, S., Sardellitti, I., Caldwell, D.: The learning-based control strategy for safe human-robot interaction exploiting task and robot redundancies. In: International Conference on Intelligent Robots and Systems, pp. 249–254 (2010)
25. Mortari, D.: On the rigid rotation concept in n-dimensional spaces. J. Astronaut. Sci. **49**(3), 401–420 (2001)

Progressive Automation of Periodic Movements

Fotios Dimeas(✉), Theodora Kastritsi, Dimitris Papageorgiou, and Zoe Doulgeri

Automation and Robotics Laboratory,
Department of Electrical and Computer Engineering,
Aristotle University of Thessaloniki, Thessaloniki, Greece
dimeasf@ee.auth.gr, tkastrit@ece.auth.gr, {dimpapag,doulgeri}@eng.auth.gr

Abstract. This paper presents the extension of the progressive automation framework for periodic movements, where an operator kinesthetically demonstrates a movement and the robotic manipulator progressively takes the lead until it is able to execute the task autonomously. The basic frequency of the periodic movement in the operational space is determined using adaptive frequency oscillators with Fourier approximation. The multi-dimensionality issue of the demonstrated movement is handled by using a common canonical system and the attractor landscape is learned online with periodic Dynamic Movement Primitives. Based on the robot's tracking error and the operator's applied force, we continuously adjust the adaptation rate of the frequency and the waveform learning during the demonstration, as well as the target stiffness of the robot, while progressive automation is achieved. In this way, we enable the operator to intervene and demonstrate either small modifications or entirely new tasks and seamless transition between guided and autonomous operation of the robot, without distinguishing among a learning and a reproduction phase. The proposed method is verified experimentally with an operator demonstrating periodic tasks in the free-space and in contact with the environment for wiping a surface.

1 Introduction

Progressive automation is a framework introduced by the authors in [3], that allows an operator to kinesthetically demonstrate repetitive tasks to a robot for seamless transition of the latter from manual robot guidance to autonomous operation. In [3] the operator demonstrates a task a few times and a variable impedance controller gradually increases the stiffness according to the correspondence between consecutive demonstrations so that the robot accurately tracks the trajectory produced by the motion generation system. Although the tasks are repetitive, they are encoded by joining discrete movement segments. The segmentation of the task into discrete movements is practical in several applications (e.g. pick and place), where the end-points of the segments are associated with the operator's input, such as signaling to the robot to open/close a gripper

F. Ferraguti et al. (Eds.): HFR 2019, SPAR 12, pp. 58–72, 2020.
https://doi.org/10.1007/978-3-030-42026-0_5

[2,10]. Although such repetitive movements can be considered periodic, there are other tasks that involve rhythmic movements and need not be segmented, such as the wiping of a surface [5] or the gait of humanoid robots [16].

To encode and determine online the basic frequency and the waveform of a periodic movement, a two layer system was proposed by Gams et al. [7]. The basic frequency is assumed to be the lowest frequency of an input signal, that is appropriate to include one task period. The first layer (Canonical System) of this method uses a number of nonlinear oscillators that adapt to the different frequency components of the input signal and the second layer (Output System) is based on periodic Dynamic Movement Primitives (DMP), which have the ability to encode periodic patterns [9]. By separating the frequency and the waveform learning, there is an advantage of independent temporal and spatial adjustment respectively. A modified approach to the first layer for determining the basic frequency of the input signal, uses a single oscillator with Fourier series approximation [20]. With Fourier approximation, there is no need to extract the basic frequency among the oscillators as in [7], which is a considerable benefit. When the objective of a task is to individually encode multiple degrees of freedom (DOF) that have coupled frequencies, the learned frequencies in each DOF might lead to drift during the reproduction phase because they might not be equal or exact multiples of each other. Another approach for multiple DOF is to use a common Canonical System for both learning and reproduction, with a common frequency. This case usually requires treatments such as logical operations, rounding of frequencies or addition of the input signals from the different DOF, which can lead to side-effects like cancelling or doubling of frequencies [7].

For learning a periodic movement that also consists of a transient part [5], such as the wiping of a surface, the authors in [4] initially segmented the wiping demonstration into the two parts and in the second phase they adapted the learned periodic DMP to apply a predefined force to the surface. Adaptation to the learned periodic movement primitives for modifying the trajectory of the robot was proposed in [8] with respect to the input from the operator (e.g. force or gestures). With this method the robot's movement could gradually adapt to the operator's coaching after multiple iterations, but under the assumption that the environment cannot change rapidly. Similarly, the authors in [14] used a passivity based iterative learning approach to gradually modify the goal of a periodic DMP with a pre-specified pattern, with respect to external forces due to changes of the environment. With the same objective, the authors in [13] proposed an adaptation mechanism to modify the spatial parameters of dynamical systems in periodic and repetitive tasks, but by having predefined motion patterns. These methods aim either to gradually adapt the learned pattern after multiple iteration or to modify the parameters of a certain pattern. As a result, they cannot handle cases in which the operator desires to significantly change the motion pattern or the frequency. Another characteristic of the aforementioned literature is the distinction between the learning and the reproduction phase. In the spirit of progressive automation, a transition between these phases should occur seamlessly, bidirectionally and without interruption [3]. Although a seamless adaptation was proposed for reshaping the task by user interaction in [13], their method assumes an encoded task prior to adaptation as opposed to ours. A seamless transition was also proposed in [19], which considered DMP learning and adaptive frequency oscillators in a single degree of freedom and gradually increased the stiffness of the impedance controller based on the input from EMG sensors attached to the operator. In that way the robot could take over the task when the predefined level of fatigue was reached. In a related approach [18], the robot switched unidirectionally from learning to autonomous execution once a predefined tracking error was reached.

In this paper we propose a method for progressive automation of periodic movements by kinesthetic guidance, extending our previous work [3] that was focused on discrete motion segments. The method utilizes adaptive frequency oscillators and periodic movement primitives for learning the frequency and waveform during the demonstration. A bidirectional seamless transition between learning and autonomous operation of the robot is achieved with the use of a role allocation strategy than can adjust the robot's stiffness and, therefore, the level of automation. This level is adjusted based on the operator's applied force and the agreement between the motion learned by the robot and the user demonstration. The contribution of this work lies in a novel modification method of the learning rules of the frequency oscillators of [20] and of the movement primitives [9], which is based on the automation level. The main advantage of this approach is that it enables autonomous execution of periodic movements through teaching by demonstration, without distinguishing between a learning and a reproduction phase. With the proposed method the learned parameters can be adjusted either for small or for significant task modifications from the operator, even by intervening during the autonomous execution and without requiring external sensors such as EMG. The effectiveness of the proposed method for fast and seamless progressive automation is verified experimentally for periodic tasks without and with contact with the environment, such as wiping of a surface.

2 Progressive Automation of Periodic Movements

This section presents the proposed methodology and is structured as follows. An overview of the system structure is initially presented Sect. 2.1 describing the key variables and the method's sub-components; these are the role allocation strategy that adjust the automation level of the robot described in Sect. 2.2, the adaptive frequency oscillators to determine the task frequency given in Sect. 2.3, the periodic DMP that encode the waveform of the demonstration presented in Sect. 2.4, and a variable stiffness controller presented in Sect. 2.5.

2.1 System Structure

Let $\mathbf{p} \in \mathbb{R}^m$ be the task coordinates of a robotic manipulator under impedance control, as shown in the block diagram of Fig. 1. At the beginning of the demonstration, the target stiffness of the robot is zero in order to allow kinesthetic guidance. To continuously adjust the role of the robot between kinesthetic guidance and autonomous operation, the target stiffness is adapted according to a role allocation law, based on the operator's force $\mathbf{F}_h$ and the tracking error of the robot. Without the system having any prior knowledge of the task, the desired trajectory $\mathbf{p}_d \in \mathbb{R}^m$ of the robot is learned incrementally during the demonstration and -simultaneously- is being provided as the reference to the impedance controller. With this approach we do not distinguish between a learning and reproduction phase. Instead, we propose a gradual increase of the target stiffness while the reference trajectory $\mathbf{p}_d$ approximates the demonstrated trajectory $\mathbf{p}$ and the operator does not apply significant forces to the robot. A decrease of the stiffness can also occur to re-enable kinesthetic guidance and allow modifications of the learned task though the application of corrective forces to the robot.

The user kinesthetically demonstrates a periodic movement to the robot and the movement is encoded by periodic DMP with incremental regression learning in each coordinate. An adaptive frequency oscillator in each coordinate i determines the basic frequency ω_i of the input signal p_i. Under the assumption that a periodic signal is available in all demonstrated degrees of freedom, then the basic frequency $\Omega \in \mathbb{R}$ of the task can be extracted as the minimum among the components:

$$\Omega = \min\{\omega_1, ..., \omega_m\}. \tag{1}$$

Synchronization of the produced trajectory $\mathbf{p}_d$ generated by the m periodic DMP, is achieved by having a common basic frequency Ω.

The concept of the proposed system involves the operator demonstrating a periodic task to the robot multiple times until the system has learned the basic frequency Ω, the phase of the periodic movement $\Phi \in \mathbb{R}$ and the desired trajectory $\mathbf{p}_d$. These estimates are updated continuously aiming to reduce the tracking error $\tilde{\mathbf{p}} = \mathbf{p} - \mathbf{p}_d$. Within this paper we only consider movement in the translational coordinates of the end-effector ($m \leq 3$), with the orientation being fixed. While the system learns the demonstrated task, the target stiffness increases and the robot gradually obtains the leading role, which is determined by the variable $\kappa \in [0, 1]$, denoting the automation level. When $\kappa = 0$, the robot can be passively guided kinesthetically with zero stiffness. While $0 < \kappa < 1$, the role is shared between the human and the robot. When $\kappa = 1$, the stiffness of the robot is maximum and it can autonomously execute the task, so it does no longer require further adaptation. For that purpose, we use the term $(1 - \kappa)$ as a weight in the adaptation rules to suspend the adaptation when the robot has learned the task. The user can intervene at any time while the robot moves autonomously -causing κ to decrease- and either modify the task (spatially or temporally) or demonstrate an entirely new task. In the following subsections we present each module of the proposed system in detail.

2.2 Role Allocation Strategy

The automation level κ transitions the role of the robot from passively following the user's demonstrations to accurately following the reference trajectory produced by the DMP. The variable Cartesian stiffness $\mathbf{K}_v$ of the robot is:

$$\mathbf{K}_v = \kappa(t) k_{max} \mathbf{I}_{m \times m} \tag{2}$$

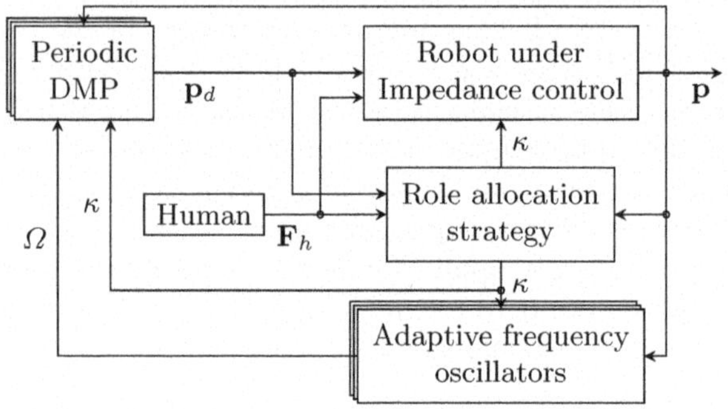

Fig. 1. Block diagram of the proposed system.

where $k_{max} \in \mathbb{R}^{>0}$ is the maximum desired stiffness for autonomous operation. The rate of change κ_r of the automation level κ depends on the external interaction force $\mathbf{F}_h$, on the tracking error $\widetilde{\mathbf{p}}$ and on the current value of $\kappa(t)$. It was originally introduced in our previous work [3] and is given here for completeness:

$$\dot{\kappa} = \begin{cases} \max\{\kappa_r, 0\}, & \kappa = 0 \\ \kappa_r, & 0 < \kappa < 1 \\ \min\{\kappa_r, 0\}, & \kappa = 1 \end{cases}, \qquad \text{with } \kappa(0) = 0, \tag{3}$$

$$\kappa_r = (\frac{\kappa}{f_r} + f_{min}) \left(1 - \left(\frac{||\mathbf{F}_h||}{\lambda_2}\right)^3 - \frac{||\widetilde{\mathbf{p}}||}{\lambda_1}\right). \tag{4}$$

The design parameters λ_1, λ_2 affect the impact of the tracking error and of the interaction force respectively on the rate of change. We utilize a power of 3 of the force magnitude to highlight the importance of the interaction force on role revoking beyond a threshold, reflected on the choice of λ_2. For example, when the robot moves autonomously with almost zero tracking error, the application of an interaction force such that $||\mathbf{F}_h|| > \lambda_2$ will revoke the leading role back to the operator. The parameter f_{min} is a positive constant to induce a gain increase when $\kappa = 0$, and f_r is a scaling term.

While the level κ increases, the robot is gradually allocated with the leading role. The transition rate κ_r depends on the current value of κ. This rate is initially slow, requiring the user to demonstrate a few periods of the movement until the frequency and waveform is learned. The rate increases with the increase of κ. When the robot moves autonomously ($\kappa = 1$), a high interaction force reverts the leading role back to the human for allowing modifications. The loss of passivity that occurs because of the variable stiffness matrix is handled with energy tanks as it is discussed in Sect. 2.5.

2.3 Adaptive Frequency Oscillators

To learn the basic frequency of the demonstrated movement, we utilize the adaptive frequency oscillators with Fourier approximation, proposed in [20]. An objective of progressive automation involves the ability of the robot to continue executing the demonstrated task autonomously after it has been encoded sufficiently. Another objective involves the ability of the operator to intervene during the autonomous execution and demonstrate any spatial or temporal modification. Regarding the temporal modification, the frequency adaptations need to stop and restart accordingly. To avoid using exteroceptive sensors such as EMG [19] and to allow continuous and bidirectional transition, in contrast to unidirectional switching as in [18], we propose the modification of the adaptation rules according to the automation level κ.

The proposed adaptation is weighted by $(1-\kappa)$ to smoothly stop the learning when the robot is in the autonomous mode $\kappa = 1$ and to smoothly re-enable it when the operator intervenes to make modifications. The oscillators are then structured as:

$$\dot{\boldsymbol{\phi}} = (1-\kappa)(\boldsymbol{\omega} - a\,\mathbf{E}\,\sin(\boldsymbol{\phi})), \tag{5}$$

$$\dot{\boldsymbol{\omega}} = -(1-\kappa)\,a\,\mathbf{E}\,\sin(\boldsymbol{\phi}), \tag{6}$$

$$\mathbf{E} = \text{diag}\left(\mathbf{p} - \hat{\mathbf{p}}\right), \tag{7}$$

where $\boldsymbol{\omega} \in \mathbb{R}^m$ is the vector of frequencies, $\boldsymbol{\phi} \in \mathbb{R}^m$ is the vector of the corresponding phases, $a \in \mathbb{R}$ is a coupling constant and $\mathbf{E} \in \mathbb{R}^{m\times m}$ is a diagonal matrix with the error between the input signal $\mathbf{p}$ and the estimate $\hat{\mathbf{p}} \in \mathbb{R}^m$. The vector of estimates $\hat{\mathbf{p}} = [\hat{p}_1, ..., \hat{p}_m]^T$ is given by:

$$\hat{p}_i = \sum_{c=0}^{M} (\alpha_{i,c}\text{cos}(\text{c}\phi_\text{i}) + \beta_{i,c}\text{sin}(\text{c}\phi_\text{i})), \quad i = 1, 2, ..., m, \tag{8}$$

where the parameter M is the number of Fourier components. The amplitudes $\alpha_{i,c}, \beta_{i,c}$ are updated according to the following rule:

$$\dot{\alpha}_{i,c} = (1-\kappa)\,\eta\,\cos(c\phi_i)\,e_i, \tag{9}$$

$$\dot{\beta}_{i,c} = (1-\kappa)\,\eta\,\sin(c\phi_i)\,e_i, \tag{10}$$

where η is the learning constant and the error e_i is the ith diagonal element of $\mathbf{E}$. Notice that the adaptation rate of the parameters in (5), (6), (9), (10) is reduced while the automation level increases ($\kappa \to 1$) and completely stops when the robot is fully autonomous ($\kappa = 1$). An intervention by the operator thought the application of an external force/torque causes the automation level to drop and the adaptation of the oscillators to be re-enabled.

Parameters a, η determine the speed of convergence and the parameter M determines the accuracy of the approximation. A sinusoidal input signal can be accurately approximated with $M = 1$. A more complex periodic signal requires more components. If we encode the complex input signal from a demonstration using less components than required, the system will learn only the strongest frequency component which is sufficient in our case, since the proposed system requires only the basic frequency. On the other hand, when more components than required are utilized, some will either converge to the same frequency or to zero frequency [20].

In the simple case when the demonstrated movement is a circle on a plane ($m = 2$), the vector $\boldsymbol{\omega}$ will converge to approximately equal frequencies. However, a slightly more complex demonstration such as the shape "8" on a plane, as the one shown in Fig. 2, consists of two frequencies where one of them is twice the value of the other. By selecting the minimum frequency for both of them according to (1), the value of Ω can be used as a common clock for the motion generation system in each coordinate. However, this can affect the accuracy of the approximated trajectory, but as we demonstrate in the experiments, this is not a problem when the frequency difference is not high. Moreover, this problem can be resolved by utilizing more basis functions in the periodic DMP that we present in the following subsection.

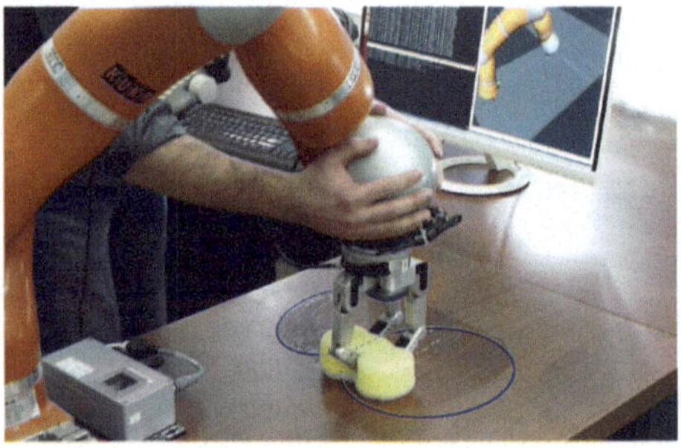

Fig. 2. Demonstrating kinesthetically a periodic wiping movement of shape "8" on a surface.

2.4 Periodic DMP

In parallel with the learning of the frequency, the waveform of the demonstrated trajectory is also learned with periodic DMP which produce the reference trajectory $\mathbf{p}_d$ to the impedance controller. This trajectory is specified by an attractor landscape towards an anchor point $\mathbf{g} \in \mathbb{R}^m$ and is governed by the phase $\Phi \in \mathbb{R}$ that is the common canonical system among the coordinates, given by:

$$\dot{\Phi} = \Omega, \quad \text{with } \Phi(0) = 0. \tag{11}$$

The reference trajectory is produced by the periodic DMP according to:

$$\dot{\mathbf{z}} = \Omega \left(a_y(\beta_y(\mathbf{g} - \mathbf{p}_d) - \mathbf{z}) + r \frac{\sum_{j=1}^{N} \mathbf{w}_j \Psi_j(\Phi)}{\sum_{j=1}^{N} \Psi_j(\Phi)} \right) \tag{12}$$

$$\dot{\mathbf{p}}_d = \Omega \mathbf{z}, \tag{13}$$

where a_y, β_y are constants and r is the amplitude control parameter which in this system is set to 1. Parameter N is the number of basis functions Ψ that are defined by:

$$\Psi_j(\Phi) = \exp(h(cos(\Phi - c_j) - 1)), \quad \forall\, j = 1, 2, ..., \; N \tag{14}$$

where h, c_j are the width and centres of the basis functions over a period.

The weights in matrix $\mathbf{w} \in \mathbb{R}^{m \times N}$ are updated online with recursive least squares. Each weight vector $\mathbf{w}_j \in \mathbb{R}^m$ of the basis function Ψ_j is updated with the Recursive Least Squares algorithm (RLS) [15]:

$$\mathbf{w}_j(t+1) = \mathbf{w}_j(t) + \Psi_j \;\mathrm{diag}(\mathbf{P}_j(t+1))\; \mathbf{e}_{r_j}(t), \tag{15}$$

where $\mathbf{P}_j \in \mathbb{R}^m$ is the vector of inverse covariance associated to the weights $\mathbf{w}_j$ with a forgetting factor λ and with elements of $\mathbf{P}_j$ $(i = 1, 2, ..., m)$:

$$P_{i,j}(t+1) = \frac{1}{\lambda}\left(P_{i,j}(t) - \frac{P_{i,j}(t)^2}{\frac{\lambda}{\Psi_j} + P_{i,j}(t)}\right). \tag{16}$$

The aim of the RLS is to minimize the error $\mathbf{e}_r$, which in this work is weighted by $(1 - \kappa)$ to smoothly stop the learning when the robot is in the autonomous mode and to smoothly re-enable it when the operator intervenes to make modifications:

$$\mathbf{e}_{r_j}(t) = (1 - \kappa)(\mathbf{f}_d(t) - \mathbf{w}_j(t)), \tag{17}$$

where $\mathbf{f}_d$ is the target trajectory shape:

$$\mathbf{f}_d = \frac{\ddot{\mathbf{p}}}{\Omega^2} - a_y(\beta_y(\mathbf{g} - \mathbf{p}) - \frac{\dot{\mathbf{p}}}{\Omega}). \tag{18}$$

Notice in (17) that when the automation level is $\kappa = 1$, the error $\mathbf{e}_r$ is zero and the adaptation of the DMP weights stops.

2.5 The Proposed Controller for Progressive Automation

To achieve progressive automation with variable robot stiffness, the following Cartesian impedance is achieved, considering a n-dof manipulator with gravity compensation, that imposes a desired Cartesian stiffness $\mathbf{K}_d \in \mathbb{R}^{6\times 6}$ and damping $\mathbf{D}_d \in \mathbb{R}^{6\times 6}$ without reshaping the inertia:

$$\mathbf{\Lambda}_x(\mathbf{q})\dot{\widetilde{\mathbf{v}}} + (\mathbf{C}_x(\mathbf{q}, \dot{\mathbf{q}}) + \mathbf{D}_d)\,\widetilde{\mathbf{v}} + \mathbf{K}_d\widetilde{\mathbf{x}} = \mathbf{F}_h, \tag{19}$$

where $\mathbf{F}_h \in \mathbb{R}^6$ is the external force from the operator, $\mathbf{q}, \dot{\mathbf{q}} \in \mathbb{R}^n$ are the robot joint position and velocity, $\widetilde{\mathbf{x}} \in \mathbb{R}^6$ is the generalized tracking error described as $\widetilde{\mathbf{x}} = [\widetilde{\mathbf{p}}^T \; \epsilon_e{}^T]^T$ with $\epsilon_e = \mathrm{vec}(\widetilde{\mathbf{Q}})$, where $\widetilde{\mathbf{Q}} = \mathbf{Q} * \mathbf{Q}_d^{-1}$ is the quaternion error between the desired orientation $\mathbf{Q}_d$ and the current orientation $\mathbf{Q}$ of the robot end-effector. The velocity error is described as $\tilde{\mathbf{v}} = [\dot{\tilde{\mathbf{p}}}^T \, \tilde{\boldsymbol{\omega}}^T]^T$ where $\tilde{\boldsymbol{\omega}} = \boldsymbol{\omega} - \boldsymbol{\omega}_d$. Within this paper, we consider a fixed orientation of the end-effector with a constant $\mathbf{Q}_d$ and $\boldsymbol{\omega}_d = \mathbf{0}_{3\times 1}$. By setting the desired Cartesian inertia equal to the robot's inertia $\boldsymbol{\Lambda}_x$, we avoid the direct feedback of $\mathbf{F}_h$.

For a constant desired stiffness $\mathbf{K}_d$ and $\mathbf{D}_d = \left[d_p \mathbf{I}_{3\times 3}\, \mathbf{0}; \mathbf{0}\; d_r \mathbf{I}_{3\times 3}\right]$, it can be easily proved that the system is passive with respect to the pair $\tilde{\mathbf{v}}$, $\mathbf{F}_h$ [10]. In this work we consider a constant orientation of the end-effector, hence we choose a high constant torsional stiffness (with $k_{rot}\mathbf{I}_{3\times 3}$) but a variable stiffness in the translation (2). Since $\mathbf{K}_d$ is now a variable matrix, then the system can lose its passivity property [6]. This problem can be overcome by introducing the varying stiffness $\mathbf{K}_v$ (with $m = 3$), of the translational components via a tank energy system, as proposed in [1,6,21]. The concept is that a virtual bounded energy is transferred from and to the dynamic system of the robot so that the whole system is passive. Thus, the variable stiffness applied by the progressive automation controller is defined as follows:

$$\mathbf{K}_d = \begin{bmatrix} \beta_r(p_t, v)\mathbf{K}_v & \mathbf{0}_{3\times 3} \\ \mathbf{0}_{3\times 3} & k_{rot}\mathbf{I}_{3\times 3} \end{bmatrix}, \tag{20}$$

where $v = -\tilde{\mathbf{p}}\mathbf{K}_v\dot{\tilde{\mathbf{p}}}$ and p_t is the virtual tank energy described by:

$$\dot{p}_t = d_p \frac{\mu(p_t)}{2} \dot{\tilde{\mathbf{p}}}^T \dot{\tilde{\mathbf{p}}} + \beta(p_t, v)\tilde{\mathbf{p}}^T \mathbf{K}_v \dot{\tilde{\mathbf{p}}}. \tag{21}$$

The functions $\mu(p_t)$, $\beta(p_t)$, $\beta_r(p_t)$ are defined in [10,11] and control the flow of the energy between the energy tank and the robot so as p_t belongs to $\mathcal{C}$ for all $t \geq 0$ with $\mathcal{C} = \{p_t \in \mathbb{R}^{\geq 0} : p_t \leq \overline{p}_t\}$ where $\overline{p}_t > 0$ is the upper limit of the virtual energy.

The whole system can be written in the following state space form:

$$\dot{\mathbf{s}} = \mathbf{H}(\mathbf{s}, \mathbf{F}_h),\; \mathbf{s}_0 = \mathbf{s}(0) \in \mathcal{Z} \tag{22}$$

where $\mathbf{s} = [\tilde{\mathbf{v}}^T \; \boldsymbol{\xi}^T \; p_t]^T \in \mathcal{Z}$, $\mathcal{Z} = \{\mathbf{s} : \mathbf{s} \in \mathbb{R}^6 \times (\mathbb{R}^3 \times \mathbb{S}^3) \times \mathcal{C}\}$ with $\boldsymbol{\xi} = [\tilde{\mathbf{p}}^T \; \tilde{\mathbf{Q}}^T]^T$and

$$\mathbf{H}(\mathbf{s}, \mathbf{F}_h) = \begin{bmatrix} \boldsymbol{\Lambda}_x^{-1}\left(-\left(\mathbf{C}_x + \mathbf{D}_d\right)\tilde{\mathbf{v}} - \mathbf{K}_d\tilde{\mathbf{x}} + \mathbf{F}_h\right) \\ \mathbf{J}_x\tilde{\mathbf{v}} \\ d_p \frac{\mu(p_t)}{2} \dot{\tilde{\mathbf{p}}}^T \dot{\tilde{\mathbf{p}}} + \beta(p_t, v)\tilde{\mathbf{p}}^T \mathbf{K}_v \dot{\tilde{\mathbf{p}}} \end{bmatrix} \tag{23}$$

where $\mathbf{J}_x = \text{diag}\left(\mathbf{I}_{3\times 3}\,,\; \frac{1}{2}\mathbf{J}_{\tilde{Q}}\right) \in \mathbb{R}^{7\times 6}$, with $\mathbf{J}_{\tilde{Q}}$ is the matrix which maps the angular velocity error to $\dot{\tilde{\mathbf{Q}}}$. i.e., $\dot{\tilde{\mathbf{Q}}} = \mathbf{J}_{\tilde{Q}}\tilde{\boldsymbol{\omega}}$ which is valid for $\boldsymbol{\omega}_d = \mathbf{0}$, which is true in our case as we consider a constant orientation.

Using the storage smooth function:

$$V = \frac{\tilde{\mathbf{v}}^T \boldsymbol{\Lambda}_x \tilde{\mathbf{v}}}{2} + (\mathbf{Q} - \mathbf{Q}_d)^T k_{rot}(\mathbf{Q} - \mathbf{Q}_d) + p_t, \tag{24}$$

its time derivative yields:

$$\dot{V} \leq -\widetilde{\mathbf{v}}^T \mathbf{D}_x \widetilde{\mathbf{v}} + \mathbf{F}_h^T \widetilde{\mathbf{v}} \tag{25}$$

while $\mathbf{D}_x = \text{diag}\left(\frac{d_p}{2}\mathbf{I}_{3\times3}\,,\ d_r\mathbf{I}_{3\times3}\right) \in \mathbb{R}^{6\times6}$. Hence, (23) is strictly output passive under the exertion of the user force $\mathbf{F}_h$ with respect to the output $\widetilde{\mathbf{v}}$ (see Definition 6.3 in [12]).

Notice that if the manipulator is redundant, (22) describes the Cartesian behavior but not the nullspace behavior. However, introducing an extra control signal like the one proposed in [17] the passivity in the redundant manipulator can be guaranteed.

3 Experimental Evaluation

To verify the effectiveness of the proposed method, an operator was asked to demonstrate periodic movements to a 7-DOF KUKA LWR4+ robot, as it is shown in Fig. 2. The operator demonstrated a movement until the robot was able to execute it autonomously and then modified the learned task spatially or temporally. In these experiments we only considered movement in the translation components of the end-effector by keeping a constant orientation (k_{rot} = 100 Nm/rad) and using the parameters of Table 1. The operator's force $\mathbf{F}_h$ was estimated from the robot's internal torque sensors. Two series of experiments were conducted, one with demonstration of free-space movements to evaluate the ability of the system in encoding a planar periodic movement, and another with a more practical application of progressively automating the wiping of a surface while keeping a constant normal force[1].

3.1 Free-Space Movements

In the first set of experiments, the user demonstrated periodic planar movements without considering contact with the environment. For m = 2, we chose a high stiffness for the Z direction with $K_v^Z = k_{max}$. The user initially demonstrated four different planar movements throughout this experiment. The results of the demonstrated tasks on the XY plane are depicted in Fig. 3. In particular, the robot's position $\mathbf{p}$, the DMP trajectory $\mathbf{p}_d$, the adapted basic frequency Ω, the automation level κ and the user's force $\mathbf{F}_h$ are shown in Fig. 3a–e. At the beginning of the experiment, the user started demonstrating a small circular motion of $\Omega = 2.4$ rad/s. After t = 15 s, when the DMP has successfully encoded the trajectory and the frequency oscillators have extracted the basic frequency, the automation level increases towards $\kappa = 1$ and the adaptation stops. The user then stops interacting with the robot, which continues to execute the circular motion autonomously. At t1 = 24 s the trajectory has been encoded by the system as it is shown in Fig. 3f. At approximately t = 34 s the operator starts

[1] Video of the experiment: https://youtu.be/uWM8VlM5y-A.

interacting again with the robot and applies a force in order to demonstrate a circular motion of bigger radius at another location. The application of the high force (Fig. 3e) reduces the automation level at $\kappa = 0$ and re-activates the adaptation until the system has learned the parameters for the new circular motion that is shown in Fig. 3g with $\Omega = 1.8\,\text{rad/s}$. Similarly, at t = 66 s the user intervenes again in order to increase the frequency of the circular motion to $\Omega = 2.6\,\text{rad/s}$ (Fig. 3h). Finally, at t = 96 s the user intervenes to demonstrate a shape "8" movement. While at the circular motions the frequency components are almost the same, in this motion pattern the frequency ω_2 in the Y coordinate is twice the ω_1 (Fig. 3c). Nevertheless, the shape has been successfully encoded (Fig. 3i) with the basic frequency of 1.1 rad/s after just 3 demonstrated periods. During this experiment, the state p_t of the energy tank remains below the upper level $\overline{p}_t = 10$ with $\beta_r(p_t, v) = 1$, so it is not illustrated. In the case the level of the tank overflows, the term $\beta_r(p_t, v) < 1$ will cause reduction of the target stiffness to maintain passivity.

Table 1. Parameters values

Param	Value	Param	Value	Param	Value
k_{max}	2500	α	50	N	30
f_r	1	η	1	λ	0.999
f_{min}	0.01	M	1	a_y	20
λ_1	0.02 m	λ_2	10 N	β_y	5

3.2 Force Controlled Wiping Task

In this experiment, the user's objective was to progressively automate a surface wiping task, while the robot applied a normal force to the surface with a sponge attached to its wrist, as shown in Fig. 2. To this aim, a hybrid impedance/force controller was implemented [19] by setting $K_v^Z = 0\,\text{N/m}$ in the Z direction and by adding the term $\mathbf{F}_f \in \mathbb{R}^6$ in the left part of (19), which is a PI force controller with a feed-forward:

$$\mathbf{F}_f = [0, 0, F_f^Z, 0, 0, 0]^T, \tag{26}$$

$$F_f^Z = f_{sp} + k_P(F_h^Z - f_{sp}) + k_I \int (F_h^Z - f_{sp})\mathrm{d}t, \tag{27}$$

where $k_P = 1$, $k_I = 1$ are the gains of the PI controller, $f_{sp} = 10\,\text{N}$ is the set-point for the desired normal force and F_h^Z is the reaction force along the Z direction estimated by the robot.

The results of the wiping task are presented in Fig. 4. After the contact of the robot with the environment is established at t = 2.5 s (Fig. 4f), the user starts demonstrating at t = 5 s a periodic wiping pattern on the surface. The proposed method achieves progressive automation within approximately 15 s, having learned the frequency and the waveform of the pattern successfully (Fig. 4c,g).

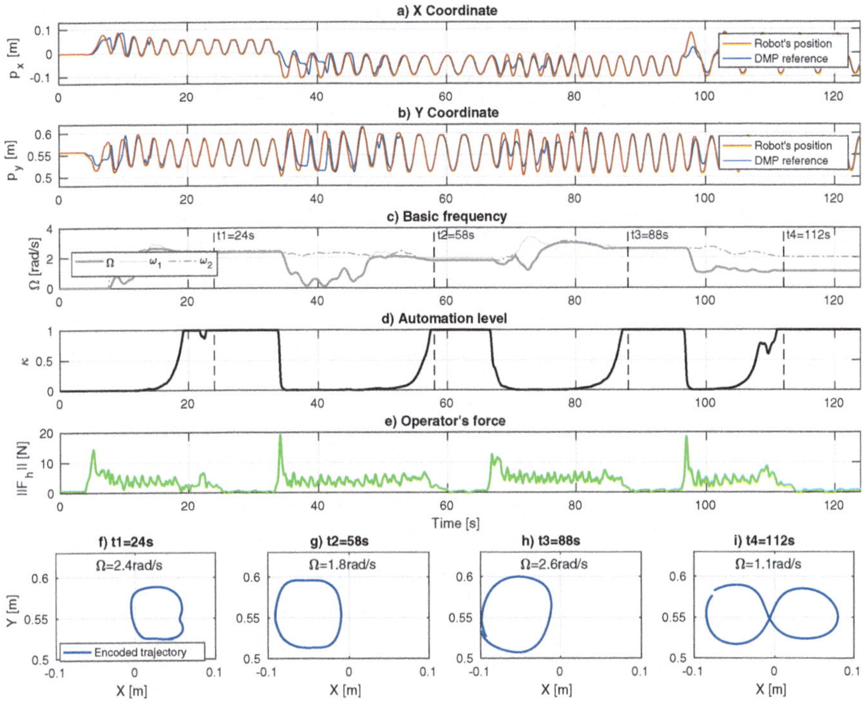

Fig. 3. Experimental results of the proposed method where a user demonstrates periodic movements in 2D and then makes modification.

The robot is at maximum stiffness after t = 18 s (with $\kappa = 1$) and the user stops interacting with the robot at 20 s (Fig. 4e). Then, the robot continues executing the task autonomously maintaining the desired force f_{sp}.

Notice that in this experiment, only the XY components of the external force $\mathbf{F}_h$ (Fig. 4e) and of the tracking error $\widetilde{\mathbf{p}}$ are considered for the role allocation strategy of Eq. (4). While the robot moves autonomously and the user has stopped interacting with it, non-zero forces appear in the XY components of $\mathbf{F}_h$ because of the friction between the sponge and the surface (shown in Fig. 4e for t > 20 s). To prevent these disturbances and the tracking errors they produce from reducing the automation level κ, the parameters λ_2, λ_1 need to be set higher than the values of the disturbances respectively. Also, notice that the proposed method assumes knowledge of the surface orientation to correctly define the position and force subspaces.

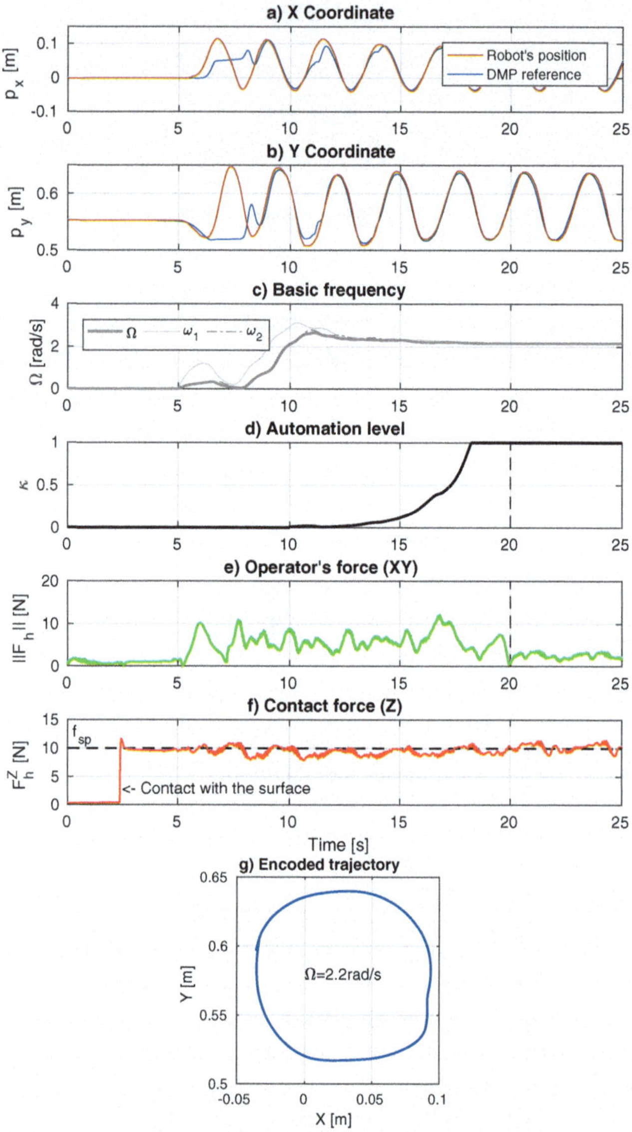

Fig. 4. Progressive automation of wiping on a surface while keeping a desired contact force.

4 Conclusions

In this work we have proposed the progressive automation of periodic movements, extending our previous work. Adaptive frequency oscillators and periodic DMP are combined with a role allocation strategy to adjust the adaptation rate and to seamlessly transition between kinesthetic robot guidance and autonomous operation. The proposed method is verified experimentally with an operator demonstrating periodic tasks in the free-space and in contact with the environment. In all of the experiments the robot was able to learn and autonomously execute a task very quickly, within 15 s to 20 s of demonstration. The user was able to intuitively determine when the robot had reached maximum stiffness and could very easily intervene to make modifications. Future work includes the achievement of progressive automation under uncertain surface orientation with a thorough passivity analysis of the hybrid controller and the ability to learn tasks that involve transitions between periodic and discrete movements.

Acknowledgement. This research is implemented through the Operational Program "Human Resources Development, Education and Lifelong Learning" and is co-financed by the European Union (European Social Fund) and Greek national funds.

References

1. Billard, A.G., Calinon, S., Dillmann, R.: Learning from humans. In: Springer Handbook of Robotics, pp. 1995–2014. Springer (2016)
2. Caccavale, R., Saveriano, M., Finzi, A., Lee, D.: Kinesthetic teaching and attentional supervision of structured tasks in human-robot interaction. Auton. Robots **43**, 1291–1307 (2018)
3. Dimeas, F., Fotiadis, F., Papageorgiou, D., Sidiropoulos, A., Doulgeri, Z.: Towards progressive automation of repetitive tasks through physical human-robot interaction. In: Ficuciello, F., Ruggiero, F., Finzi, A. (eds.) Human Friendly Robotics. Springer Proceedings in Advanced Robotics, vol. 7, pp. 151–163. Springer, Cham (2019)
4. Do, M., Schill, J., Ernesti, J., Asfour, T.: Learn to wipe: a case study of structural bootstrapping from sensorimotor experience. In: 2014 IEEE International Conference on Robotics and Automation (ICRA), pp. 1858–1864. IEEE, May 2014
5. Ernesti, J., Righetti, L., Do, M., Asfour, T., Schaal, S.: Encoding of periodic and their transient motions by a single dynamic movement primitive. In: IEEE-RAS International Conference on Humanoid Robots, pp. 57–64 (2012)
6. Ferraguti, F., Secchi, C., Fantuzzi, C.: A tank-based approach to impedance control with variable stiffness. In: 2013 IEEE International Conference on Robotics and Automation, pp. 4948–4953, May 2013
7. Gams, A., Ijspeert, A.J., Schaal, S., Lenarčič, J.: On-line learning and modulation of periodic movements with nonlinear dynamical systems. Auton. Robots **27**(1), 3–23 (2009)
8. Gams, A., Petrič, T., Do, M., Nemec, B., Morimoto, J., Asfour, T., Ude, A.: Adaptation and coaching of periodic motion primitives through physical and visual interaction. Robot. Auton. Syst. **75**, 340–351 (2016)
9. Ijspeert, A.J., Nakanishi, J., Schaal, S.: Learning attractor landscapes for learning motor primitives. Advances in Neural Information Processing Systems, pp. 1547–1554 (2002)
10. Kastritsi, T., Dimeas, F., Doulgeri, Z.: Progressive automation with DMP synchronization and variable stiffness control. IEEE Robot. Autom. Lett. **3**(4), 3789–3796 (2018)

11. Kastritsi, T., Sidiropoulos, A., Doulgeri, Z.: A pHRI framework for modifying a robot's kinematic behaviour via varying stiffness and dynamical system synchronization. In: 26th Mediterranean Conference on Control and Automation (MED), pp. 33–38 (2018)
12. Khalil, H.: Nonlinear Systems, 3rd edn. Prentice Hall, Upper Saddle Rive (2002)
13. Khoramshahi, M., Laurens, A., Triquet, T., Billard, A.G.: From human physical interaction to online motion adaptation using parameterized dynamical systems. In: 2018 IEEE/RSJ International Conference on Intelligent Robots and Systems, pp. 1361–1366 (2018)
14. Kramberger, A., Shahriari, E., Gams, A., Nemec, B., Ude, A., Haddadin, S.: Passivity based iterative learning of admittance-coupled dynamic movement primitives for interaction with changing environments. In: 2018 IEEE/RSJ International Conference on Intelligent Robots and Systems (IROS), pp. 6023–6028. IEEE, October 2018
15. Ljung, L., Söderström, T.: Theory and Practice of Recursive Identification. Electrical Engineering. MIT Press, Cambridge (1983)
16. Nakanishi, J., Morimoto, J., Endo, G., Cheng, G., Schaal, S., Kawato, M.: Learning from demonstration and adaptation of biped locomotion. Robot. Auton. Syst. **47**(2–3), 79–91 (2004)
17. Ott, C., Kugi, A., Nakamura, Y.: Resolving the problem of non-integrability of nullspace velocities for compliance control of redundant manipulators by using semi-definite Lyapunov functions. In: 2008 IEEE International Conference on Robotics and Automation, pp. 1999–2004. IEEE (2008)
18. Peternel, L., Petrič, T., Oztop, E., Babič, J.: Teaching robots to cooperate with humans in dynamic manipulation tasks based on multi-modal human-in-the-loop approach. Auton. Robots **36**(1–2), 123–136 (2014)
19. Peternel, L., Tsagarakis, N., Caldwell, D., Ajoudani, A.: Robot adaptation to human physical fatigue in human-robot co-manipulation. Auton. Robots **42**(5), 1011–1021 (2018)
20. Petrič, T., Gams, A., Ijspeert, A.J., Žlajpah, L.: On-line frequency adaptation and movement imitation for rhythmic robotic tasks. Int. J. Robot. Res. **30**(14), 1775–1788 (2011)
21. Landi, C.T., Ferraguti, F., Fantuzzi, C., Secchi, C.: A passivity-based strategy for coaching in human-robot interaction. In: 2018 IEEE International Conference on Robotics and Automation (ICRA), pp. 3279–3284. IEEE (2018)

Fault-Tolerant Physical Human-Robot Interaction via Stiffness Adaptation of Elastic Actuators

Florian Stuhlenmiller[1], Rodrigo J. Velasco-Guillen[2(✉)], Stephan Rinderknecht[1], and Philipp Beckerle[1,2]

[1] Institute for Mechatronic Systems in Mechanical Engineering, Technical University Darmstadt, Darmstadt, Germany
[2] Elastic Lightweight Robotics Group, Robotics Research Institute, Technical University Dortmund, Dortmund, Germany
rodrigo.velasco@tu-dortmund.de

Abstract. Elastic actuators are popular in human-robot interaction as they can improve human safety and efficiency. Yet, such actuators are more complex than rigid ones and might be subject to additional technical faults, e.g., stiffness changes. This paper extends previous studies on stiffness-fault-tolerant physical human-robot interaction (pHRI) through control adaptation, introducing new methods for stiffness estimation and fault evaluation. Kalman filters with different measurement signals and system models estimating the actual stiffness value of the elastic element are compared. Faults are evaluated by analyzing the structural durability and compensated by adapting an impedance controller to provide a desired interaction stiffness. Experiments with a series elastic actuator underline the feasibility of the evaluation and compensation methods for attaining safe and reliable pHRI. Results show that stiffness estimation during pHRI is possible when the actuator friction and interaction torque is either negligible or well known, or when the torque at the spring is measured.

Keywords: Physical human-robot interaction · Elastic actuator · Fault evaluation · Fault-tolerant control

1 Introduction

Physical human-robot interaction (pHRI) has become of great importance in robotics and due to this, safety and dependability are requirements for robot design and control [1]. Accordingly, elastic robotic actuators receive strong interest as they support human safety through allowing deformation in case of unintentional contact [2,3]. Furthermore, elastic actuator designs can reduce energy consumption by adapting the natural dynamics of the system to a desired motion [4,5]. However, the introduction of elasticity increases the system complexity and can lead to a more frequent occurrence of faults [6]. A fault management system is then important, particularly for pHRI, to ensure safety and reliability [7].

F. Ferraguti et al. (Eds.): HFR 2019, SPAR 12, pp. 73–87, 2020.
https://doi.org/10.1007/978-3-030-42026-0_6

Over the years, different elastic actuator designs have been proposed with fixed or variable elasticity [8]. Series elastic actuators (SEA) are the most common, a concept where drive and load are coupled by an elastic element [9]. Elastic actuators with variable stiffness are referred to as variable stiffness actuators (VSAs) and can adapt their energy efficiency and interaction behavior through varying their compliance [5,10].

According to [11], a fault is defined as an unpermitted deviation of at least one characteristic property or feature of the system from the acceptable condition, i.e., usually the standard condition. When dealing with elastic elements, changes in stiffness can be related to the structural durability of the material. Low and slow stiffness reduction can occur due to cyclic loading of a spring, while large and quick stiffness reduction might be due to misuse loadings or special events that could cause plastic deformation or fractures in extreme cases.

Technical faults in elastic actuators are of practical relevance [6] and thus fault-tolerant control strategies are required to obtain a safe and reliable system. This means that faults should be compensated in a way that avoids system failures [11], which can be obtained by designing a control system capable of adapting to faults [12]. Analytical methods for fault-tolerant control, such as process model-based approaches, provide deeper system insight than those which are constrained to analyzing signals [11,12]. However, they generally require sufficient knowledge of the system and fault impact [11], which is usually obtained from mathematical models, heuristically, or by data-driven approaches, e.g., machine learning. With process model-based approaches, fault-tolerant control strategies can be designed either actively by redesigning the control law at the occurrence of a fault, passively by making the control system robust, or a combination of both [12].

This paper extends previous studies on fault-tolerant pHRI through stiffness estimation and control adaptation in elastic actuators [13], introducing new methods for stiffness estimation using Kalman Filters. Experiments on a variable torsional stiffness (VTS) actuator present the feasibility of the methods to contribute to attaining safer and more reliable pHRI. Section 2 presents the actuator, its model, and the behavior of faulty elastic elements. Section 3 describes the fault evaluation method based on structural durability and a fault-tolerant control strategy to compensate for changes in stiffness. Experimental results are presented and discussed in Sect. 4 and the study is concluded in Sect. 5.

2 Actuator and Fault Behavior

This section presents a variable torsional stiffness (VTS) actuator [14,15] as an exemplary experimental SEA. The model of the system is presented as well as the behavior of stiffness faults in its elastic element.

2.1 Actuator Model

The VTS actuator is shown in Fig. 1. It contains a torsional spring that connects actuator 1 and the load, i.e., a pendulum. A counter bearing, moved by actuator 2,

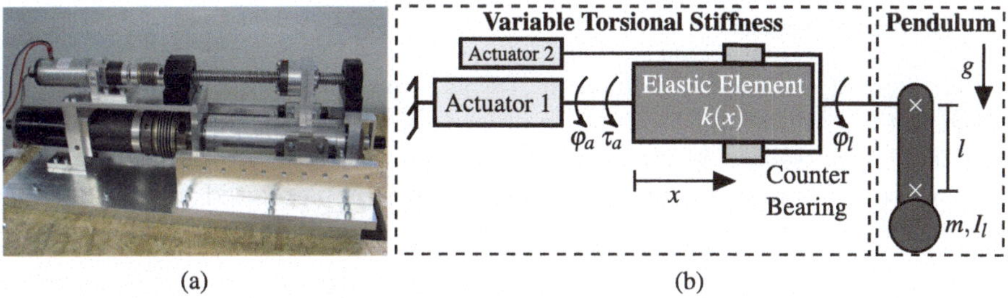

Fig. 1. Photography (left) and structure (right) of the VTS-actuator

adjusts the active length of the spring and thus controls the stiffness k of the elastic element, making it possible to emulate stiffness faults by modifying the stiffness value.

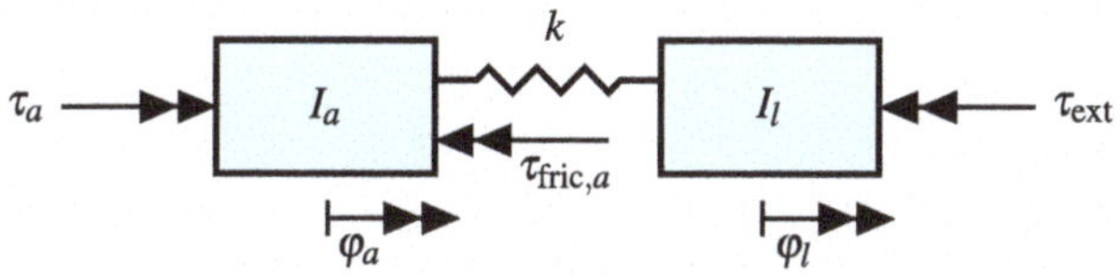

Fig. 2. Two-degree-of-freedom model of a SEA with friction

The two-degree-of-freedom model of such an SEA is shown in Fig. 2, which includes a friction torque $\tau_{\text{fric},a}$ acting on the actuator side. The serial stiffness k couples the actuator inertia I_a and link inertia I_l. The actuator, with position φ_a, exerts the torque τ_a and the link, with position φ_l, is loaded by the external torque τ_{ext}. Based on the model presented in [16] and neglecting damping and centrifugal effects, the following equations represent the system model:

$$\begin{bmatrix} I_l & 0 \\ 0 & I_a \end{bmatrix} \begin{bmatrix} \ddot{\varphi}_l \\ \ddot{\varphi}_a \end{bmatrix} + \begin{bmatrix} k & -k \\ -k & k \end{bmatrix} \begin{bmatrix} \varphi_l \\ \varphi_a \end{bmatrix} = \begin{bmatrix} -\tau_{\text{ext}} \\ \tau_a - \tau_{\text{fric},a} \end{bmatrix}. \tag{1}$$

The external load on the VTS-actuator consists of the gravitational torque exerted by the pendulum and the interaction torque τ_{int}. With mass m, distance to the center of rotation l, and gravitational acceleration g, the external torque is defined as:

$$\tau_{\text{ext}} = mgl \sin(\varphi_l) + \tau_{\text{int}}. \tag{2}$$

The combined Coulomb-tanh and viscous friction model [17] can be used to characterize the friction torque:

$$\tau_{\text{fric},a} = \tau_c \tanh(S\dot{\varphi}_a) + f_v\dot{\varphi}_a, \tag{3}$$

where τ_c is the Coulomb friction torque, S is a coefficient that determines the shape of the tanh function modeling Coulomb friction, and f_v is the viscous friction coefficient. While S is set to a fixed value, τ_c and f_v can be estimated online.

2.2 Stiffness Faults

The concept of damage is defined in [18] as a property which diminishes the strength of a material until failure. In order to quantify this concept, a damage variable D is presented in [18], which measures the structural changes caused by microscopic damage in the material. This variable is related to the effective stress $\tilde{\sigma}$ acting on the damaged cross-section that resists forces:

$$\tilde{\sigma} = \frac{\sigma}{1 - D}, \tag{4}$$

where σ is the Cauchy stress tensor.

The damage variable D can then be evaluated by its influence on the elasticity, such that $\tilde{E} = E\,(1 - D)$ represents the effective Young modulus, with E being the modulus of the undamaged material [18]. Thus, if the modulus E of a material is known and $\tilde{E}$ is measured once the material is damaged, the damage variable can be calculated as:

$$D = 1 - \tilde{E}/E. \tag{5}$$

In this way, the elasticity of the elastic element can be used to determine the structural durability of the material, with the damage variable going from $D = 0$ (no damage) to $D = 1$ (no structural durability). However, this method is limited to uniformly distributed damage in the strained material volume [18].

Given the proportional relationship between Young modulus and stiffness, and assuming that the relation between the cross-section area and length of the element does not change in damage conditions, the damage variable can be evaluated as:

$$D = 1 - \tilde{k}/k_0, \tag{6}$$

where $\tilde{k}$ is the effective stiffness measured for the damaged material and k_0 the stiffness of the undamaged material.

With respect to pHRI, structural durability information can be used to determine whether a stiffness-fault-tolerant controller [13] could compensate for the stiffness change or operation should be terminated. In this sense, faults can be categorized in hazard classes, which defines fault evaluation [11].

3 Fault Evaluation and Compensation

This section presents methods for evaluating and compensating for stiffness faults based on different approaches of online stiffness estimation. Through online compensation, the desired pHRI characteristics are maintained even in faulty states.

3.1 Stiffness Estimation

In order to evaluate and compensate for stiffness faults during operation, a stiffness estimate $\bar{k}$ has to be determined online. To this end, the method presented in [19] features the model-based determination and low-pass filtering of the torque in the elastic element from either the link-side or the actuator side from Eq. (1). This residual signal is passed to a recursive least squares algorithm with fixed directional forgetting [13] to dynamically compute $\bar{k}$. However, stiffness estimation based on link-side information requires measurement of τ_{int} [19]. In contrast, the actuator-side is independent from τ_{int} according to Eqs. (1) and (2), but the friction torque $\tau_{\text{fric},a}$ has to be well-kown for accurate determination of the stiffness. For the VTS, experimental stiffness estimation with the method from [13] was not adequate for fault compensation due to uncertainties in non-linear friction characteristics. Therefore, the extended Kalman filter (EKF) and the unscented Kalman filter (UKF) are examined in this work. While EKFs are widely used for estimation, they rely on linearizations which introduce approximation errors or computational complexity [20,21]. In contrast, UKFs overcome these approximation issues by utilizing the unscented transformation method [20].

In this paper, different implementations of EKF and UKF are compared: EKF utilizing the complete actuator model presented in Eq. (1) and both an EKF and UKF with a torque sensor measuring the spring torque τ_s. For each method, a discrete model is defined, such as:

$$\mathbf{x}_{i+1} = \mathbf{F}(\mathbf{x}_i, \mathbf{u}_i) + \mathbf{w}_i, \qquad \mathbf{y}_i = \mathbf{H}(\mathbf{x}_i) + \mathbf{v}_i. \tag{7}$$

System states, parameters, and/or disturbances are included in the state vector $\mathbf{x}$. Known or measured control inputs are presented as $\mathbf{u}$, measurements as $\mathbf{y}$, the non-linear system model as $\mathbf{F}(\cdot)$, and the non-linear observation model as $\mathbf{H}(\cdot)$. Process and measurement noise are represented as $\mathbf{w}_i$ and $\mathbf{v}_i$, respectively. The state is initialized by an initial estimation $\bar{\mathbf{x}}_0 = \mathrm{E}(\mathbf{x}_0)$ and an initial covariance $\mathbf{P}_0 = \mathrm{E}(\mathbf{x}_0\mathbf{x}_0^{\mathrm{T}})$.

For EKF, a continuous model can be discretized using finite differences as $\mathbf{F}(\mathbf{x}_i, \mathbf{u}_i) = \mathbf{x}_i + T_0\dot{\mathbf{x}}_i(\mathbf{x}_i, \mathbf{u}_i)$, where T_0 represents the time step. The linearization of the system and measurements are captured in the Jacobian matrices $\mathbf{A}$ and $\mathbf{C}$, such that:

$$\mathbf{A}_i = \left.\frac{\partial \mathbf{F}}{\partial \mathbf{x}}\right|_{\bar{\mathbf{x}}_i}, \qquad \mathbf{C}_i = \left.\frac{\partial \mathbf{H}}{\partial \mathbf{x}}\right|_{\bar{\mathbf{x}}_i} \tag{8}$$

The EKF method applies a time update to predict the state vector $\bar{\mathbf{x}}_{i+1}^-$ with covariance $\mathbf{P}_{i+1}^-$, and a measurement update to correct the estimation and calculate $\bar{\mathbf{x}}_{i+1}$ with covariance $\mathbf{P}_{i+1}$. The equations of the method are the following:

$$\begin{aligned} \bar{\mathbf{x}}_{i+1}^- &= \mathbf{F}(\bar{\mathbf{x}}_i, \mathbf{u}_i) & \mathbf{K}_{i+1} &= \mathbf{P}_{i+1}^- \mathbf{C}_i^{\mathrm{T}} (\mathbf{C}_i \mathbf{P}_{i+1}^- \mathbf{C}_i^{\mathrm{T}} + \mathbf{R})^{-1} \\ \mathbf{P}_{i+1}^- &= \mathbf{A}_i \mathbf{P}_i \mathbf{A}_i^{\mathrm{T}} + \mathbf{Q} & \bar{\mathbf{x}}_{i+1} &= \bar{\mathbf{x}}_{i+1}^- + \mathbf{K}_{i+1}(\mathbf{y}_{i+1} - \mathbf{H}(\bar{\mathbf{x}}_{i+1}^-)) \\ & & \mathbf{P}_{i+1} &= (\mathbf{I} - \mathbf{K}_{i+1}\mathbf{C}_i)\mathbf{P}_{i+1}^-, \end{aligned}$$

where $\mathbf{K}$ is the Kalman gain, while $\mathbf{Q}$ and $\mathbf{R}$ are the covariance matrices for process and measurement noise, respectively.

The UKF, on the other hand, does not require explicit calculation of Jacobians [21]. Instead, the state vector is redefined as the concatenation of the original state vector and noise variables as $\mathbf{x}_i^a = \left[\mathbf{x}_i^{\mathrm{T}}\ \mathbf{w}_i^{\mathrm{T}}\ \mathbf{v}_i^{\mathrm{T}}\right]^{\mathrm{T}}$. The state and covariance of the new state vector can be initialized as:

$$\bar{\mathbf{x}}_0^a = \left[\bar{\mathbf{x}}_0^{\mathrm{T}}\ \mathbf{0}\ \mathbf{0}\right]^{\mathrm{T}}, \qquad \mathbf{P}_0^a = \operatorname{diag}\left(\mathbf{P}_0, \mathbf{Q}, \mathbf{R}\right). \tag{9}$$

The UKF applies a time update to predict the state vector $\bar{\mathbf{x}}_{i+1}^-$ with covariance $\mathbf{P}_{i+1}^-$, and the measurement $\bar{\mathbf{y}}_{i+1}^-$ with covariance $\mathbf{P}_{i+1}^{\mathrm{YY}}$. Followed by a measurement update, where the estimated state $\bar{\mathbf{x}}_{i+1}$ and its covariance $\mathbf{P}_{i+1}$ are computed. The method makes use of tunable parameters: α determines the spread of sigma points, κ is a scaling factor, and β incorporates prior knowledge of the distribution. The complete equations of the UKF are given in [21] and [22].

EKF with Complete Actuator Model: Using the complete model in Eq. (1) allows to estimate the stiffness k, as well as the Coulomb friction torque τ_c and the viscous friction coefficient f_v. For this, the measured torques in the actuator τ_a and interaction τ_{int} are used as inputs and the positions of link φ_l and actuator φ_a are the measurements:

$$\mathbf{x} = \left[\varphi_l\ \dot{\varphi}_l\ \varphi_a\ \dot{\varphi}_a\ k\ \tau_c\ f_v\right]^{\mathrm{T}}, \quad \mathbf{y} = \left[\varphi_l\ \varphi_a\right]^{\mathrm{T}}, \quad \mathbf{u} = \left[\tau_a\ \tau_{\text{int}}\right]^{\mathrm{T}},$$

$$\mathbf{F}(\mathbf{x}_i, \mathbf{u}_i) = \begin{bmatrix} \varphi_l \\ \dot{\varphi}_l \\ \varphi_a \\ \dot{\varphi}_a \\ k \\ \tau_c \\ f_v \end{bmatrix}_i + T_0 \begin{bmatrix} \dot{\varphi}_l \\ \frac{1}{I_l}\left(-mgl\sin\varphi_l - k\left(\varphi_l - \varphi_a\right) - \tau_{\text{int}}\right) \\ \dot{\varphi}_a \\ \frac{1}{I_a}\left(\tau_a + k\left(\varphi_l - \varphi_a\right) - \tau_c \tanh\left(S\dot{\varphi}_a\right) - f_v\dot{\varphi}_a\right) \\ 0 \\ 0 \\ 0 \end{bmatrix}_i .$$

EKF with Torque Sensor at the Spring: This method uses the torque measurement at the spring τ_s as input, and the integration of the deflection $\int_i^{i+1}\left(\varphi_l - \varphi_a\right)\mathrm{d}t$ as measurement. The method is defined as:

$$\mathbf{x} = \left[\int_i^{i+1}\left(\varphi_l - \varphi_a\right)\mathrm{d}t\ k\right]^{\mathrm{T}}, \quad \mathbf{y} = \left[\int_i^{i+1}\left(\varphi_l - \varphi_a\right)\mathrm{d}t\right], \quad \mathbf{u} = \left[\tau_s\right],$$

$$\mathbf{F}\left(\mathbf{x}_i, \mathbf{u}_i\right) = \begin{bmatrix} \int_{i-1}^{i}\left(\varphi_l - \varphi_a\right)\mathrm{d}t \\ k_i \end{bmatrix} + T_0 \begin{bmatrix} \frac{\tau_s}{k} \\ 0 \end{bmatrix}_i .$$

Notice that without integrating the deflection, the system model $\mathbf{F}(\cdot)$ would require the numerical differentiation of the noisy sensor signal τ_s to calculate the corresponding Jacobian. Furthermore, the integration also ensures persistent excitation for static motions, which is not possible when $\dot{\tau}_s = 0$. However, a difficulty might occur as the integration may increase to infinity.

UKF with Torque Sensor at the Spring: Using the torque measurement at the spring τ_s as input, measuring the position difference $\varphi_l - \varphi_a$ and making use of the relation $\varphi_l - \varphi_a = \tau_s/k$, this method is defined as such:

$$\mathbf{x} = [k], \quad \mathbf{y} = [\varphi_l - \varphi_a], \quad \mathbf{u} = [\tau_s], \quad \mathbf{H}(\mathbf{x}_i) = \left[\frac{\tau_s}{k}\right]_i, \quad \mathbf{F}(\mathbf{x}_i, u_i) = [k_i]$$

3.2 Fault Evaluation

In the presence of stiffness faults the structural durability of the elastic element can be determined in order to ensure the integrity of the element for its usage. As mentioned in Sect. 2.2, the damage variable D can be calculated using the effective stiffness. Applying the estimated stiffness value $\bar{k}$ as the effective stiffness in Eq. (6), an estimated damage variable $\bar{D}$ can be determined as:

$$\bar{D} = 1 - \bar{k}/k_0. \tag{10}$$

Modeling the elastic element as a torsional rod, which applies to the example of VTS [15,23], the cross-section typical maximum shear stress $\sigma_{\max}$ can be calculated:

$$\sigma_{\max} = \frac{\tau_s}{c_r b^3}, \tag{11}$$

where c_r is the torsional resistance factor and b is the thickness of the element.

Applying the estimated damage variable and cross-section shear stress to Eq. (4), an estimated effective stress $\bar{\tilde{\sigma}}$ can be determined as:

$$\bar{\tilde{\sigma}} = \frac{\sigma_{\max}}{1 - \bar{D}} = \frac{\tau_s}{c_r b^3 (1 - \bar{D})}. \tag{12}$$

Thus, the effective shear stress $\bar{\tilde{\sigma}}$ can be estimated making use of the measured torque at the spring τ_s and the estimated damage variable $\bar{D}$. Notice that as the damage variable increases, so does the effective stress. A threshold needs to be set for $\tilde{\sigma}$ such that the motion can be performed without breaking the spring. As long as this threshold is not violated, compensating for stiffness faults to ensure safe pHRI via control appears applicable.

The critical effective shear stress $\tilde{\sigma}_{\text{critial}}$ that would allow a safe interaction can be determined applying the von Mises criterion [24], using the maximum tensile strength of the elastic element σ_y and a notch factor K_t, such that:

$$\tilde{\sigma}_{\text{critical}} = \frac{\sigma_y}{\sqrt{3} K_t}. \tag{13}$$

This is used as a threshold with which faults are evaluated: If $|\bar{\tilde{\sigma}}| > \tilde{\sigma}_{\text{critical}}$, the structural durability of the element is not acceptable and thus there is a critical fault.

3.3 Fault Compensation

The stiffness-fault-tolerant control strategy proposed in [13] relies on a passivity-based controller [25], where the actuator torque is determined from:

$$\tau_a = \frac{I_a}{I_{a,d}} u + \left(1 - \frac{I_a}{I_{a,d}}\right) k(\varphi_a - \varphi_l). \tag{14}$$

To achieve passivity of the system [25], the control input u is defined as:

$$u = I_{a,d}\ddot{\varphi}_{a,d} + k(\varphi_{a,d} - \varphi_{l,d}) + k_c(\varphi_{a,d} - \varphi_a) + d_c(\dot{\varphi}_{a,d} - \dot{\varphi}_a) \tag{15}$$

Given a desired link position $\varphi_{l,d}$, the desired actuator position $\varphi_{a,d}$ can be computed via inverse dynamics from Eq. (1):

$$\varphi_{a,d} = \varphi_{l,d} + \frac{1}{k}\left(mgl \sin\left(\varphi_{l,d}\right) + I_l \ddot{\varphi}_{l,d}\right) \tag{16}$$

The first two terms of Eq. (15) represent load compensation for the desired link and actuator motion, respectively. The last two terms represent PD-control of the actuator, which introduces a virtual stiffness k_c and viscous damping d_c, and specifies the disturbance reaction of the controlled system.

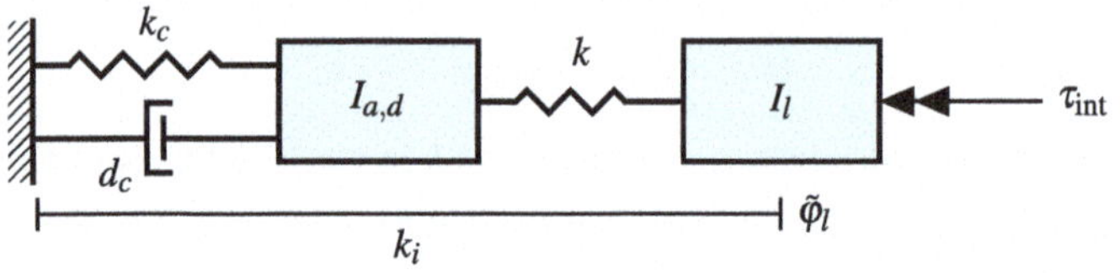

Fig. 3. Characteristic structure of impedance-controlled SEA

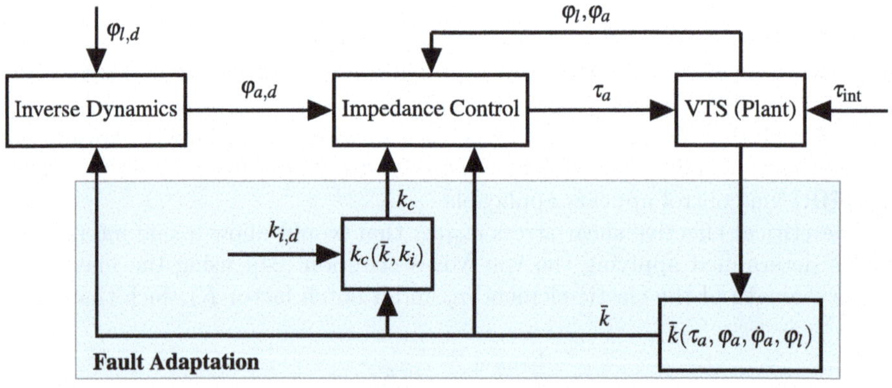

Fig. 4. Block diagram of the fault-tolerant control strategy for dependable pHRI

Figure 3 shows the characteristic structure of the impedance-controlled SEA. The system dynamics can now be defined with respect to the position error $\tilde{\varphi}$:

$$\begin{bmatrix} I_l & 0 \\ 0 & I_{a,d} \end{bmatrix} \begin{bmatrix} \ddot{\tilde{\varphi}}_l \\ \ddot{\tilde{\varphi}}_a \end{bmatrix} + \begin{bmatrix} 0 & 0 \\ 0 & d_c \end{bmatrix} \begin{bmatrix} \dot{\tilde{\varphi}}_l \\ \dot{\tilde{\varphi}}_a \end{bmatrix} + \begin{bmatrix} k & -k \\ -k & k + k_c \end{bmatrix} \begin{bmatrix} \tilde{\varphi}_l \\ \tilde{\varphi}_a \end{bmatrix} = \begin{bmatrix} \tau_{\text{int}} \\ 0 \end{bmatrix} \tag{17}$$

The interaction stiffness k_i is a series configuration of virtual and physical stiffness, defined as the relation between the interaction torque τ_{int} and link position error $\tilde{\varphi}_l$:

$$k_i = \frac{k k_c}{k + k_c} \quad \rightarrow \quad k_c(\bar{k}, k_i) = \frac{\bar{k} k_i}{\bar{k} - k_i}. \tag{18}$$

Fault compensation makes use of the estimated stiffness $\bar{k}$ as feedback for inverse dynamics, the impedance control law, and to calculate the virtual stiffness k_c. The stiffness parameter k_c is adapted to attain a desired interaction stiffness $k_{i,d}$, which ensures fault-tolerant pHRI [13]. The control strategy is presented in Fig. 4.

Regarding stability, $k_c > 0$ is required to achieve a passive and thus stable system according to [25]. Applying this condition to Eq. (18), one can derive that $k_i < k$ is necessary. Hence, within the stability limits, the pHRI characteristics can only be softer than the physical stiffness.

4 Experimental Results

This section presents experimental results obtained with the VTS actuator exhibiting the parameters given in Table 1. Stiffness estimation with the different methods described in Sect. 3.1 is compared considering constant and sinusoidal trajectories. Furthermore, results of fault evaluation and stiffness fault compensation are presented with and without interaction. For all experiments, encoder measurements are available for position of actuator φ_a and link φ_l, as well as torque sensor measurement for τ_s. The actuator torque τ_a is known from the control output. Furthermore, τ_{int} is calculated from the measured spring torque and link-side dynamics extracted from Eq. (1).

Table 1. Parameters of the VTS actuator

Description	Value	Description	Value
Inertia of actuator I_a	$1.15\,\text{kg m}^2$	Torsional stiffness k	$50 \ldots 350\,\text{N m rad}^{-1}$
Inertia of pendulum I_l	$0.9\,\text{kg m}^2$	Spring thickness b	$0.014\,\text{m}$
Mass of pendulum m	$6.81\,\text{kg}$	Resistance factor c_r	0.208
Length of pendulum l	$0.36\,\text{m}$	Max. tensile strength σ_y	55 MPa
Gravitational acceleration g	$9.81\,\text{m s}^{-2}$	Notch factor K_t	1.3
Coulomb constant S	100		

4.1 Stiffness Estimation

One can expect that the degradation of the stiffness of an elastic element happens slowly over its lifespan. By controlling the stiffness in the VTS actuator it is possible to emulate an accelerated hypothetical degradation over a short period of time. With an initial value $k_0 = 175\,\mathrm{N\,m\,rad^{-1}}$, the stiffness is repeatedly reduced by $25\,\mathrm{N\,m\,rad^{-1}}$ at times 40, 60, 80, and 100 s. The control strategy presented in Sect. 3.3 is activated on all experiments to compensate for the stiffness changes using the EKF with torque sensor at the spring as the stiffness estimation method.

To tune the Kalman filters, it is crucial to determine the process and measurement noise covariance matrices $\mathbf{Q}$ and $\mathbf{R}$ which result in the desired performance [26]. Filter parameters are tuned manually to balance convergence speed for quick fault compensation and smoothness of the signal to avoid introducing additional noise into the control algorithm. Additionally, for the UKF the parameters $\alpha = 1 \times 10^{-3}$, $\kappa = 0$ and $\beta = 2$ are utilized, as recommended in [21] for Gaussian distributions. The tuned covariance matrices used for the estimation methods are shown in Appendix A.

Figure 5 shows the results obtained from an experiment where the link holds a constant desired position of 10° with no interaction present. Figure 5(a) shows how the actuator position (green) increases every time a stiffness change occurs, while the link position (blue) remains relatively constant. This is due to the stiffness fault compensation which results in a variation of the actuator position to maintain the desired link position and pHRI characteristics. Figure 5(b) shows the comparison of the different stiffness estimation methods. Notice that

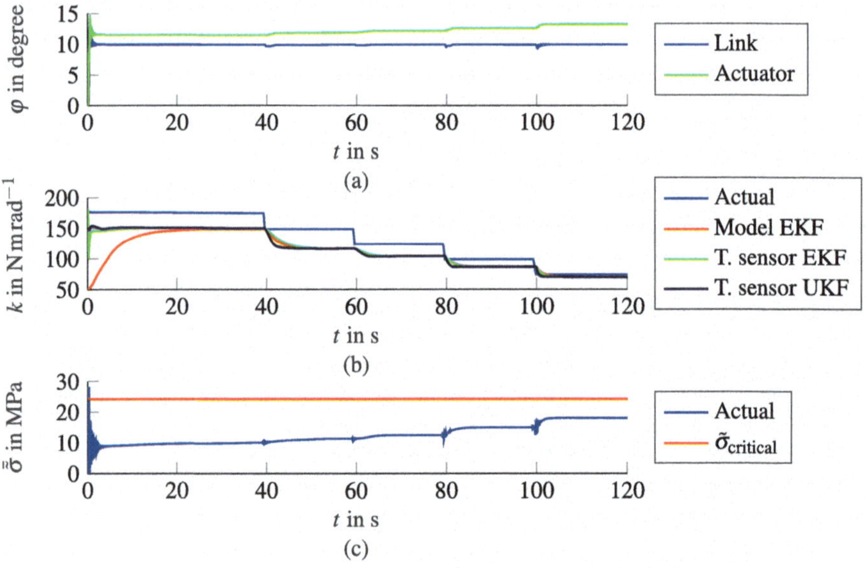

Fig. 5. Progression of 10° constant position experiment

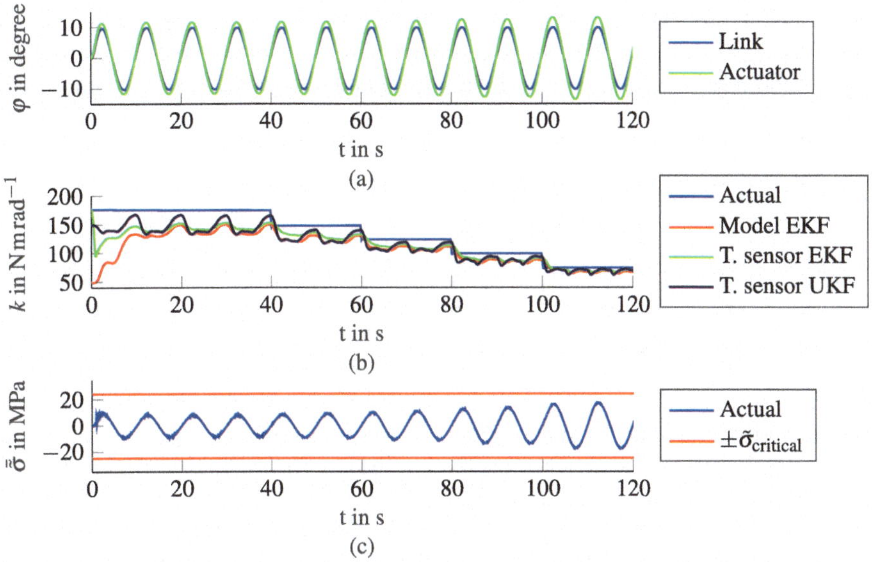

Fig. 6. Progression of 0.1 Hz oscillation experiment

all stiffness estimations (red, green, and black) are lower than the experimental evaluation (blue). Yet, they all eventually converge. Although all estimation methods show similar behavior, estimation with the UKF using a torque sensor at the spring (black) yielded the fastest response. Figure 5(c) shows the hypothetical effective shear stress applied to the elastic element. For the structural durability analysis, the stiffness is estimated using the EKF with torque sensor method. Notice how $\bar{\tilde{\sigma}}$ increases as the stiffness degrades, this occurs due to the increment of the macroscopic damage variable $\bar{D}$ as the estimated stiffness $\bar{k}$ decreases. Since $\bar{\tilde{\sigma}}$ never surpasses the critical shear stress $\tilde{\sigma}_{\text{critical}}$, the elastic element remains with an acceptable structural durability throughout the experiment and thus the emulated faults are not considered critical.

Figure 6 shows the results obtained from an experiment where the desired link position is an oscillatory motion with a frequency of 0.1 Hz and an amplitude of 20°. Similarly to the previous experiment, the link position (green) remains with a constant amplitude while the amplitude of the actuator position (blue) increases as the stiffness change is compensated by the controller as shown in Fig. 6(a). The comparison of the stiffness estimation methods in Fig. 6(b) shows that estimated stiffness (red, green, and black) is lower than the actual physical value (blue) for all methods. While all estimations converge, frequency-dependent variations of estimated stiffness occur. Notice that although all estimation methods show a similar behavior, estimation with EKF using a torque sensor at the spring (green) yielded the smallest variations. The hypothetical effective shear stress is presented in Fig. 6(c) and outlines an amplitude increase of $\bar{\tilde{\sigma}}$ with degrading stiffness, once again estimated using the EKF with torque sensor method. Since the magnitude $\bar{\tilde{\sigma}}$ never surpasses the critical shear stress $\tilde{\sigma}_{\text{critical}}$, the elastic element remains with an acceptable structural durability throughout the experiment and thus the emulated faults are not considered critical.

4.2 Fault-Tolerant Interaction

Figure 7 shows the results obtained from an experiment where the link holds a constant desired position of $-10°$. Faults reducing stiffness repeatedly by $25\,\text{N}\,\text{mrad}^{-1}$ are emulated at 40, 60, 80, and 100 s as in previous experiments. Additionally, interaction with a human occurs for approximately 2 s at 17, 52, 69, 92, and 107 s. Figure 7(a) shows the effect of interaction on the position of link (blue) and actuator (green). Fault compensation is again visible as the actuator position decrements every time a stiffness change occurs. Figure 7(b) compares the examined stiffness estimation methods. Similar to previous experiments, all estimated stiffness values are lower that the actual physical value (blue). The estimations of all methods show sudden deviations when interaction is present, with the estimation from the EKF using the model (red) being most robust due to the knowledge of the interaction torque τ_{int}. A modified version of the EKF

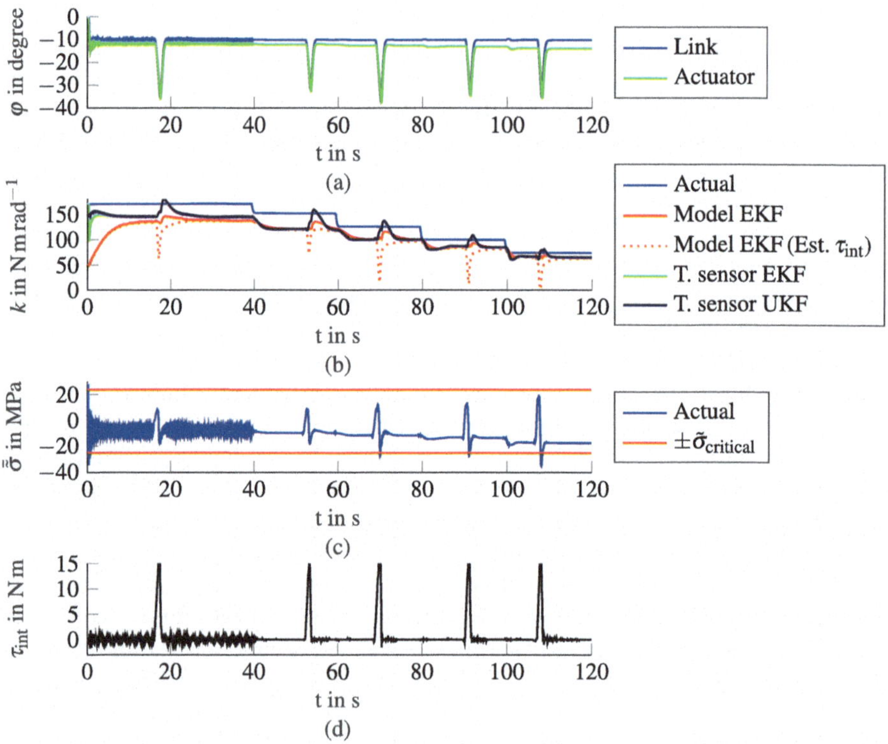

Fig. 7. Progression of $-10°$ constant position experiment with interaction

using the complete actuator model is presented (dotted red), in which τ_{int} is added to the state vector $\mathbf{x}$ and removed from the control input $\mathbf{u}$, meaning that the interaction torque is estimated and not measured. Notice how the lack of τ_{int} measurement leads to a less accurate stiffness estimation at every interaction. Thus, properly estimating stiffness and interaction at the same time appears not to be possible. The hypothetical effective shear stress on the elastic element shown in Fig. 7(c) shows increment in $\bar{\bar{\sigma}}$ with degrading stiffness as in the previous experiments. Yet, the shear stress threshold is violated at times 69, 92, and 107 s, which would indicate an imminent failure. Those violations appear to be related to interaction rather than the emulated stiffness faults. The occurring interaction torques between human and system are displayed in Fig. 7(d) and do not vary in amplitude despite changing stiffness due to fault-tolerant control.

5 Conclusions

This paper presents methods to evaluate and compensate stiffness faults of serial elastic actuators to provide safe and reliable physical human-robot interaction (pHRI). Experiments were carried out in a variable torsional stiffness actuator emulating a fast stiffness degradation of the elastic element to compare three methods for online stiffness estimation. An extended Kalman filter using a model of the actuator yielded adequate convergence of the estimated stiffness to the actual physical value. This was possible with sufficient knowledge of friction, as well as measuring the actuator and interaction torque as inputs, and using the actuator and link position signals as measurements. Extended and unscented Kalman filters using a torque sensor at the spring achieved similar adequate convergence results. Stiffness estimation with these methods is possible with spring torque measurement as input and the position deflection as measurement. Tuning of the filters allowed the extended Kalman filter to be less frequency-dependent and the unscented Kalman filter to have a faster response for constant amplitudes. It can be concluded that adequate stiffness estimation is possible when the actuator friction is negligible or well known. Otherwise, applicable stiffness estimation in presence of physical human-robot interaction requires measurement of the spring torque or additional knowledge detailing interaction torque.

Fault evaluation is based on structural durability analysis of the elastic element during the hypothetical stiffness degradation. Results show that the shear stress over the effective area of the damaged cross-section gets closer to the material limitations as the stiffness degrades. This study underlines the feasibility of the fault evaluation and control method, compensating stiffness faults while keeping acceptable structural durability. However, the emulated faults as well as the shear stress thresholds are hypothetical examples that might deviate from real conditions. Thus, in future works, deeper examination of material-specific degradation behavior is required. Moreover, more lightweight spring designs could become a possibility when such a method is mature. The fault compensation strategy based on impedance control originally presented in [13] is experimentally investigated in this work. Results show that fault compensation is successful in attaining a desired link position and maintaining desired pHRI characteristics. Finally, this work contributes into the development of safer and more reliable pHRI in various applications such as wearable or collaborative robotics.

Acknowledgment. This work was supported by a Deutsche Forschungsgemeinschaft (DFG) Research Grant (no. BE 5729/1).

A Filter Covariance Matrices

EKF with complete actuator model:

$$\mathbf{Q} = 1 \times 10^{-7}\text{diag}\,(1, 1000, 1, 1000, 100000, 100, 100)\,, \ \mathbf{R} = 1 \times 10^{-7}\mathbf{I}_7.$$

EKF with torque sensor at the spring:

$$\mathbf{Q} = \text{diag}\left(1 \times 10^{-20}, 1 \times 10^{-15}\right), \ \mathbf{R} = 1 \times 10^{-20}\mathbf{I}_2.$$

UKF with torque sensor at the spring:

$$\mathbf{Q} = \left[1 \times 10^{-5}\right], \ \mathbf{R} = \left[1 \times 10^{-8}\right].$$

References

1. Haddadin, S., Albu-Schaeffer, A., De Luca, A., Hirzinger, G.: Collision detection and reaction: a contribution to safe physical human-robot interaction. In: IEEE/RSJ International Conference on Intelligent Robots and Systems, pp. 3356–3363 (2008)
2. Haddadin, S., Albu-Schäffer, A., Hirzinger, G.: Safe physical human-robot interaction: measurements, analysis and new insights. In: ISRR, vol. 66, pp. 395–407. Springer, Heidelberg (2007)
3. Lens, T., von Stryk, O.: Investigation of safety in human-robot-interaction for a series elastic, tendon-driven robot driven robot arm. In: IEEE/RSJ International Conference on Intelligent Robots and Systems (2012)
4. Vanderborght, B., Van Ham, R., Lefeber, D., Sugar, T.G., Hollander, K.W.: Comparison of mechanical design and energy consumption of adaptable, passive-compliant actuators. Int. J. Robot. Res. **28**(1), 90–103 (2009)
5. Verstraten, T., Beckerle, P., Furnémont, R., Mathijssen, G., Vanderborght, B., Lefeber, D.: Series and parallel elastic actuation: impact of natural dynamics on power and energy consumption. Mech. Mach. Theory **102**, 232–246 (2016)
6. Beckerle, P.: Practical relevance of faults, diagnosis methods, and tolerance measures in elastically actuated robots. Control Eng. Pract. **50**, 95–100 (2016)
7. Filippini, R., Sen, S., Bicchi, A.: Toward soft robots you can depend on. IEEE Robot. Autom. Mag. **15**(3), 31–41 (2008)
8. Vanderborght, B., Albu-Schaeffer, A., Bicchi, A., Burdet, E., Caldwell, D., Carloni, R., Catalano, M., Eiberger, O., Friedl, W., Ganesh, G., Garabini, M., Grebenstein, M., Grioli, G., Haddadin, S., Hoppner, H., Jafari, A., Laffranchi, M., Lefeber, D., Petit, F., Stramigioli, S., Tsagarakis, N., Damme, M.V., Ham, R.V., Visser, L., Wolf, S.: Variable impedance actuators: a review. Robot. Auton. Syst. **61**(12), 1601–1614 (2013)

9. Pratt, G.A., Williamson, M.M.: Series elastic actuators. In: IEEE/RSJ International Conference on Intelligent Robots and Systems (1995)
10. Schiavi, R., Grioli, G., Sen, S., Bicchi, A.: VSA-II: a novel prototype of variable stiffness actuator for safe and performing robots interacting with humans. In: IEEE International Conference on Robotics and Automation (2008)
11. Isermann, R.: Fault-Diagnosis Systems: An Introduction from Fault Detection to Fault Tolerance. Springer, Heidelberg (2006)
12. Blanke, M., Kinnaert, M., Lunze, J., Staroswiecki, M.: Diagnosis and Fault-Tolerant Control. Springer, Heidelberg (2010)
13. Stuhlenmiller, F., Perner, G., Rinderknecht, S., Beckerle, P.: A stiffness-fault-tolerant control strategy for reliable physical human-robot interaction. In: Human Friendly Robotics, pp. 3–14. Springer, Heidelberg (2019)
14. Lendermann, M., Singh, B.R.P., Stuhlenmiller, F., Beckerle, P., Rinderknecht, S., Manivannan, P.V.: Comparison of passivity based impedance controllers without torque-feedback for variable stiffness actuators. In: IEEE/ASME International Conference on Advanced Intelligent Mechatronics (2015)
15. Schuy, J., Beckerle, P., Wojtusch, J., Rinderknecht, S., von Stryk, O.: Conception and evaluation of a novel variable torsion stiffness for biomechanical applications. In: IEEE/RAS & EMBS International Conference on Biomedical Robotics and Biomechatronics, pp. 713–718 (2012)
16. Spong, M.W.: Adaptive control of flexible joint manipulators. Syst. Control Lett. **13**, 15–21 (1989)
17. Andersson, S., Söderberg, A., Björklund, S.: Friction models for sliding dry, boundary and mixed lubricated contacts. Tribol. Int. **40**(4), 580–587 (2007)
18. Lemaitre, J., Dufailly, J.: Damage measurements. Eng. Fract. Mech. **28**(5), 643–661 (1987)
19. Flacco, F., De Luca, A.: Residual-based stiffness estimation in robots with flexible transmissions. In: 2011 IEEE International Conference on Robotics and Automation (ICRA), pp. 5541–5547. IEEE (2011)
20. Julier, S., Uhlmann, J.: Unscented filtering and nonlinear estimation. Proc. IEEE **92**(3), 401–422 (2004)
21. Wan, E.A., Van Der Merwe, R.: The unscented Kalman filter for nonlinear estimation. In: Adaptive Systems for Signal Processing, Communications, and Control Symposium 2000. AS-SPCC. The IEEE2000, pp. 153–158. IEEE (2000)
22. Wu, M., Smyth, A.: Application of the unscented kalman filter for real-time nonlinear structural system identification. Struct. Control Health Monit. **14**, 971–990 (2007)
23. Schuy, J., Beckerle, P., Faber, J., Wojtusch, J., Rinderknecht, S., von Stryk, O.: Dimensioning and evaluation of the elastic element in a variable torsion stiffness actuator. In: IEEE/ASME International Conference on Advanced Intelligent Mechatronics (2013)
24. de Souza Neto, E.A., Perić, D., Owen, D.R.J.: Computational Methods for Plasticity. Wiley, Hoboken (2008)
25. Ott, C.: English Cartesian Impedance Control of Redundant and Flexible-Joint Robots. Springer, Heidelberg (2008)
26. Kaur, N., Kaur, A.: A review on tuning of extended kalman filter using optimization techniques for state estimation. Int. J. Comput. Appl. **145**(15), 1–5 (2016)

Designing an Expressive Head for a Help Requesting Socially Assistive Robot

Tim van der Grinten[1(✉)], Steffen Müller[1], Martin Westhoven[2], Sascha Wischniewski[2], Andrea Scheidig[1], and Horst-Michael Gross[1]

[1] Neuroinformatics and Cognitive Robotics Lab, Ilmenau University of Technology, Ilmenau, Germany
tim.van-der-grinten@tu-ilmenau.de
[2] Unit 2.3 Human Factors, Ergonomics, Division 2 Products and Work Systems, German Federal Institute for Occupational Safety and Health, Dortmund, Germany
https://www.tu-ilmenau.de/neurob/
https://www.baua.de/EN/About-BAuA/Organisation/Division-2/Unit-2-3.html

Abstract. In this paper, we present the developments regarding an expressive robot head for our socially assistive mobile robot *HERA*, which among other things is serving as an autonomous delivery system in public buildings. One aspect of that task is contacting and interacting with unconcerned people in order get help when doors are to open or an elevator has to be used. We designed and tested a robot head comprising a pan-tilt unit, 3D-printed shells, animated eyes displayed on two LCD-screens, and three arrays of RGB-LEDs for communicating internal robot states and attracting potential helpers' interest. An online-study was performed to compare variations of eye-expression and LED lighting. Data was extracted from the answers of 139 participants. Statistical analysis showed significant differences in identification performance for our intended eye-expressions, perceived politeness, help intentions, and hedonic user experience.

Keywords: Mechatronic design · Expressive robot head · Social robots

1 Introduction

1.1 Application Scenario

The work presented here is part of the robotic research project *FRAME* (Assisted elevator use and room access for robots by involving helpers) [1], which is dealing with a fundamental problem for socially assistive robots without any manipulators [2]. If the robot has to pass through a closed door or has to ride the elevator, especially in public buildings, it can ask for help from people passing by [3]. The aim of the *FRAME* project is to use several robotic platforms in three different

This project is funded by the German Federal Ministry of Education and Research (BMBF) within FRAME (16SV7829K).

F. Ferraguti et al. (Eds.): HFR 2019, SPAR 12, pp. 88–102, 2020.
https://doi.org/10.1007/978-3-030-42026-0_7

application scenarios. One is an in-house postal delivery system, the second one is a robot for item transportation and measuring air pollutant parameters in a small factory, and the third one is a messenger application in a public hospital.

1.2 Robot

The robot platform used is a *SCITOS G5* robot by *Metralabs GmbH* named *HERA*. Figure 1 shows the robot which has a movable base with a differential drive and a swivel caster at the rear. Its battery enables the robot for autonomous operation for about 3 h, after which it can autonomously return to its charging station. For navigation purposes, the robot is equipped with several sensors. It has two SICK laser range scanners at a height of 40 cm covering 360° around the robot. For obstacle avoidance, two *ASUS Xtion* depth cameras are covering the closer environment, and an inertial measurement unit (IMU) is used for supporting the correction of the robot's odometry. For perception of people in its surroundings, the robot has three wide angle RGB-cameras covering the whole panorama as well as a *Kinect 2* RGB-D camera on a pan-tilt unit (PTU). Interaction takes place on a 10" tablet and by means of a projector for visualization of navigation goals and advises. Originally the robot had no explicit head. Only the PTU with the *Kinect 2* sensor looking around was present, and the robot lacked any kind of personality, which could be helpful when it comes to interaction with unknown helpers. Against this background, we were looking for a robotic head, that could be placed on top of the robot platform.

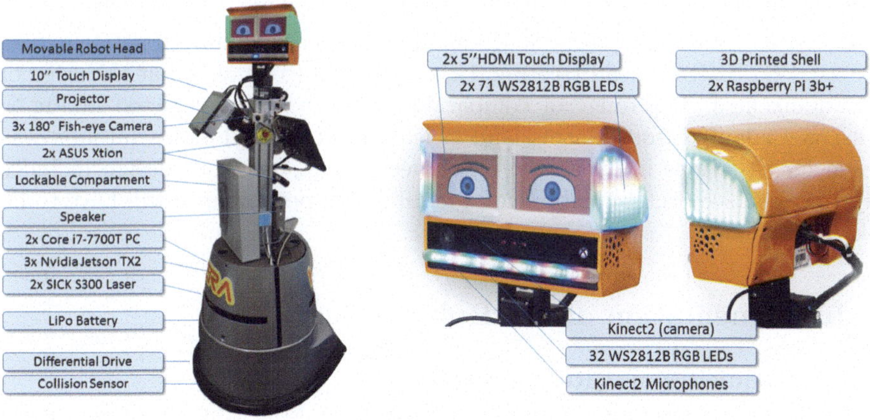

Fig. 1. Robot *HERA* with its sensors, interaction devices and the robot's head.

2 Related Work and Design Decisions

The literature provides results mostly for isolated criteria of single design features and some hints at interaction effects.

Song and Yamada [4] studied colored lights, vibrations, and sounds on a real, but simplistic robot. This served to explore the effects in general, but also to verify the intended effects of their design decisions. As Baraka and Veloso [5] summarize, lights are seldom coupled with the state of a robot, although they are sometimes used to underline emotion. In a series of studies, they consequently showed how lighting patterns can be used to clarify a robot's state.

Bennet and Šabanoviç [6] report results from a study on minimalist robot facial features for emotion expression. They use the upper and lower outline of the eyes as well as those of the mouth and achieved high accuracy in expression identification. Eyes can be a subtle cue for observation and increase cooperative behavior [7]. They can transport emotions in human-robot interaction [8] and should therefore be present when social interaction is a core part of a robot's task. As Marsh, Ambady, and Kleck [9] found, sad or fearful expressions facilitate approach behaviours in perceivers, which is helpful when asking for a favor. Lee, Šabanoviç, and Stolterman [10] performed a qualitative study on social robot design. Their participants reported that eyes should not be too far apart and also not be too detailed. Also overly large eyes were reported to be intimidating and to be inducing a feeling of surveillance.

The design of the eye part of the robot's head orients itself on 'in the wild' examples. A more comic-like style was chosen due to results regarding the uncanny valley, that is the lowering of trust and likeability when robots get more human-like, but not enough so. Following the general understanding of the uncanny valley phenomenon, Mathur and Reichling [11] reported that staying on the mechanic-looks side of the so-called mechano-humanness score range can also yield high values for trust and likeability as well as low response times.

Therefore, considering also the trade-off between economic and performance-wise aspects, HERA's head (see Fig. 1) was decided to be aimed at this area of machine-likeness. The range of comparable robot heads is depicted in Fig. 2, which was adapted from Mathur and Reichling [11]. A sad or fearful look was chosen for situations in which the robot needs help. Since our design includes only the eyes (see Sect. 3), expression identification performance will have to be checked again.

3 Implementation of the Robot Head

3.1 Prerequisites and Requirements

The design of the head pursued different goals and considered the following constraints: main reason for having a head is, that it provides the robot some personality. It is meant as a communication device for interacting with people, who can read from the eye contact, that the robot is addressing individual persons when talking. Furthermore, the head is a means for expressing different internal states of the robot. In the first instance, these are global states like normal operation, error state, or the need for help, but it can also transport emotional states during a dialog. A visual feedback signal with synthetic mouth

Fig. 2. Robotic heads ordered from machine-like to human-like as compiled by Mathur and Reichling [11] showing the target area of our design.

movements helps to associate voice outputs to the robot head, which physically come from the speaker at the robot's center. As a constructional constraint, we had to integrate the comparatively large *Kinect 2* sensor consisting of cameras, infrared boosters, and a microphone array, which must not be covered. This defines a minimum size of the head, while a maximum size and weight is defined by the used *FLIR PTU-E46-17P70T* PTU's performance (max. payload 4.5 kg).

From earlier projects [12] with a mechanical head, where eye-balls and lids are driven by individual stepper motors and mechanical transmission, we had experiences with failure susceptibility of mechanical components. It took several redesign cycles until the hardware was robust for long term operation. Therefore, the new head design should involve as few mechanical parts as possible restricting us to a display solution, where the eyes are animated. Together with the given minimum width of the head defined by the *Kinect 2* sensor, the size of the front side would be rather huge considering available display formats. This and the positive experiences from the eye design in the SERROGA project [13] led to the idea of using two smaller LCD-displays, each requiring an own HDMI source.

3.2 Hardware

Figure 1 shows the details of our robot head. Its main structure consists of four 3D-printed parts:

1. a base containing the *Kinect 2* sensor and a mounting plate for the electronics (made from ABS)
2. the back side of the base giving room for the fan and cables (ABS)
3. the glasses which give a frame for the displays and LEDs (translucent PLA)
4. an easy to remove hood covering the inside (ABS)

Fig. 3. Wiring of the *WS2812B* LEDs. Greyed out LEDs are not installed but only shown for understanding.

The shells show organic openings for ventilation and heat dissipation, which makes use of the *Kinect 2*'s built-in fan.

The two 5" touch displays are controlled by two *Raspberry Pi 3 Model B* SoCs which are connected via HDMI and SPI for the touch signal. This allows for a natural reaction when somebody touches the robot into the eye.

The aforementioned glasses frame (see Fig. 1) is additionally equipped with two arrays of 71 *WS2812B* LEDs on the two sides as shown in Fig. 3. These LED-arrays are intended for expressing the robot's state in order to attract attention. Another 32 *WS2812B* LEDs are arranged as a line in between the cameras and microphones of the *Kinect 2* sensor comprising the robot's mouth. All LEDs are controlled by one of the *Raspberry Pi 3* SoCs.

3.3 Eye Animations and Control

The robot's eyes are realized by means of the *OGRE* 3D engine. Originally designed for computer game graphics, this engine allows for rendering articulated mesh objects and supports weighted superposition of animations for individual objects that can be generated with a 3D animation software like *Blender*. Figure 4 shows the object structure used for the eyes. Each of the two *Raspberry Pi 3* SoCs renders a frontal orthogonal view of that 3D geometry.

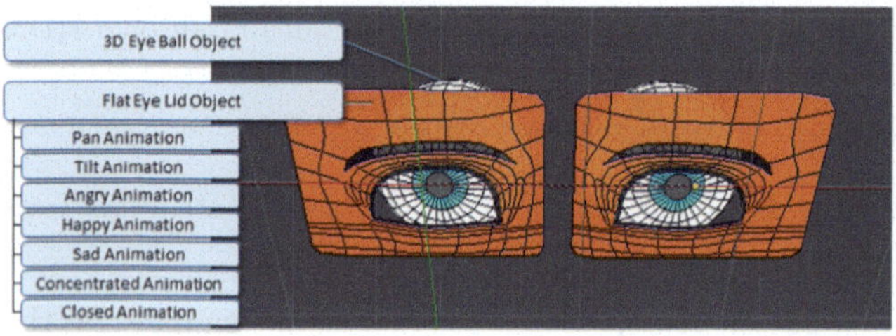

Fig. 4. Structure of the animated mesh for the robot's eyes.

Each eye consists of two 3D objects. One eye ball, that is not animated but can rotate according to the desired gaze direction (important for making eye contact between robot and user), and a flat object comprising the skin with the eye lid and the eye brows. The eye lid object is used for expressing different states but also has to consider the gaze direction and can do an eye blink animation. In the animation software, seven base animations have been defined that have to be combined in real time by means of the *OGRE* engine. The pan and tilt animations cover a movement from left to right and from top to bottom respectively. The emotional expression animation, e.g. angry, happy, sorrow or concentrated (see Fig. 5), contain a transition from a neutral state ($t = 0$) to the fully expressed emotional state ($t = 1$) over the animation's time parameter t (see Fig. 6). The close animation covers the transition from a completely opened to a closed eye.

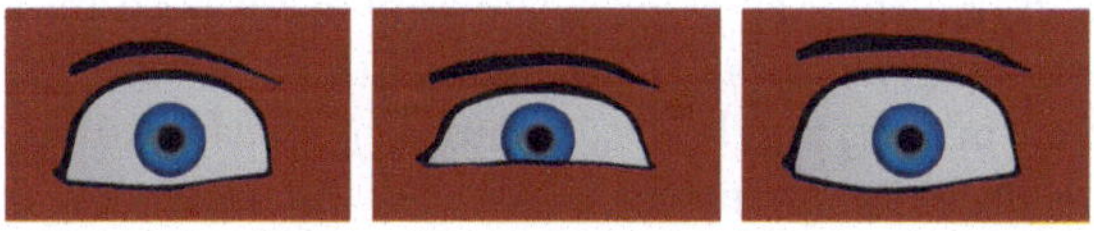

Fig. 5. Normal (left), happy (middle), angry (right) state on the robot's left eye.

For realizing the actual superposition of the base animations, three individual controllers are used in order to define the eye's appearance. Each of the controllers manipulates the time parameter t in the respective animation and provides a mixing weight w in range $[0, 1]$, which defines the influence of the individual animations in the resulting rendering.

Gaze Target Controller. This controller gets a 3D position from the application where to look at, specified relative to the robot. Considering the current angles of the PTU, the direction of gaze in head based coordinates can be determined easily, which specifies the desired pan and tilt angle of the eye balls[1].

Furthermore, these angles directly define the parameter t for the pan and tilt animations of the eye lids (position in the animation going from left to right and bottom to upward direction). The mixing weights w for these animations are set proportional to the amount of elongation. Pan and tilt angles close to the angular limits generate weights of 1 overriding the influence of other animations, while angles near zero generate low mixing weights, and therefore, the margin for expressing other animations is higher. The interplay between the direction of the eyes and the head orientation is described below.

Emotion State Controller. This is used in order to realize the smooth transition between changing emotional states, which are defined by a five element vector $E = [e_{angry}, e_{curious}, e_{happy}, e_{sorrow}, e_{concentrated}]$ comprising the amount of

[1] Head and eyes movement: https://youtu.be/PHBMrr7HQzI.

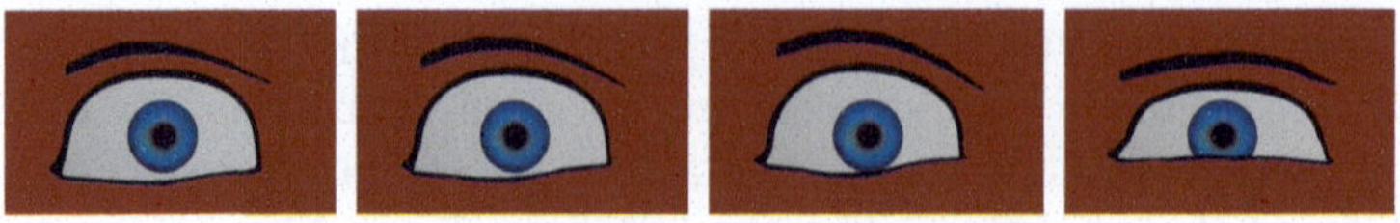

Fig. 6. Morphing between normal and happy state on the robot's left eye. Happy factor (from left to right): 0.0, 0.5, 0.75, 1.0

excitation for each of the emotional states. The application can choose arbitrary combinations of the base emotions and not necessarily has to respect a limit to the speed of change. If the application changes this vector E, the controller slowly approximates the rendered state to it by means of a recursive averaging to avoid jitter (see Fig. 6). The time constant for that is about 0.5 s.

The E values are directly used as the parameter t of the base animations. The more active an emotion, the closer the animation is to its full articulation. Additionally, the mixing weights w of the animations are activated proportional to the E values. The mixing weights w_{pan} and w_{tilt} for the pan and tilt animation of the eye lid are scaled with $1-\max_i(e_i)$. Thus, if emotions are activated, the pan and tilt animations are less visible, which prevents from diluting the articulation of strong emotions in case of high gaze direction angles.

Blink State Controller. This controller is responsible for animating the eye blink events. If triggered, the animation runs continuously from open to closed state and back (t parameter of the animation), while the mixing weights w of all the other base animations are gradually reduced to zero until the closed position is reached. The weight of the closed animation is increased instead. When opening the eyes again, the weights are faded back to the original values, such that the former state is visible again. By means of that, the blink animation can smoothly fade in from arbitrary states and reaches its fully closed position without interference with other emotional states and gazing direction[2].

A last feature of the eye displays is the adaption of external illumination distribution. From the RGB camera in the *Kinect 2* sensor, we get an overall impression of the light sources in the surroundings. The pixels in the RGB image are clustered in order to find color and intensity of the bright spots in the scene around the robot. Since the *OGRE* engine allows to define up to eight point light sources for illumination of the rendered scene, the brightest clusters are used to control the position and color of these lights in the *OGRE* scene. By means of this, it is possible to prevent scary glowing eyes in the dark, while in bright sunlight the contrast is as high as possible.

[2] Eye emotions and blink: https://youtu.be/XrsamLVvKO8.

3.4 LEDs and Control

The three arrays of installed *WS2812B* LEDs as introduced in Sect. 3.2 are handled as three different rectangular matrices internally, simply ignoring missing LEDs (see Fig. 3), which makes it easier to control them. For each of these three groups, five programmable parameters are accessible by the application:

1. light mode
2. color (RGB)
3. speed
4. direction (*UP*, *DOWN*, *FRONT*, *BACK*)
5. mode depending special parameter

The light mode represents the effect the LEDs should show. 14 different modes are available (see Table 1). Since the installed *WS2812B* LEDs are RGB, the color parameter accepts a three element vector containing 8-bit values for red, green and blue. The speed gives the cycle speed in Hz, while for asymmetric effects the direction can be specified. The last parameter can control effect-dependent settings like the duty cycle or a line width (see Table 1).

Table 1. Light modes, for visualization see (LED modes and parameters: https://youtu.be/4WF3vIgE5G4)

Light mode	Description	Special parameter
OFF	LEDs off	–
ON	LEDs on	–
BLINK	LEDs blinking	duty cycle $[0, 1]$
PULSE	LEDs pulsing	Duty cycle $[0, 1]$
RUNNING	Running line	Line width
PENDULUM	Back-and-forth running line	Line width
GROWING	Growing line	–
SHRINKING	Shrinking line	–
GROWSHRINK	First growing then shrinking line	–
RAINBOW	LEDs show rainbow color	–
RBCYCLE	Rainbow running light (red first)	Wavelength
RBCYCLE2	Rainbow running light (blue first)	Wavelength
MATRIX	Matrix effect	Probability of new dots $[0, 1]$
AUDIO	light line, width controlled by audio signal	audio signal scaling factor

4 Application

4.1 Attention Generation and Status Output

For our application (see Sect. 1), where the robot relies on the help of humans passing by, it is important for it to attract the attention of human helpers. Especially over longer distances, the easiest and most unobtrusive way for attracting attention is a visual stimulus. For this purpose, the robot can make the LEDs in its head light up or flash.

After having attracted the attention of a potential helper, the robot has to communicate with him or her. To unload the voice channel of communication, the color of the LEDs can be used to assist in displaying information about the robot to the helper. This can make it easier to understand the robot's state. Since one of the application scenarios will be in a factory setting, color codes consistent to other machinery should be used. We use the ISO standard *IEC 60204-1* [14] for designing the color design. This standard demands special colors for different machine states (see Table 2). Deducing therefrom, the robot has to light the green LEDs when it can do things on its own and the blue LEDs when it needs help. Yellow is also used to warn observers, e.g. when moving through narrow passages. Red is reserved for critical errors and thus not part of normal robot operation.

Table 2. Meaning of light color according to ISO standard *IEC 60204-1* [14]

Light color	Meaning
RED	Critical error
YELLOW	Abnormal state, imminent critical error
GREEN	Normal state
BLUE	Mandatory action required

4.2 Gaze Following

For being more social, the robot's head and eye gaze should be directed towards the human interaction partner. This can be accomplished coarsely by the installed *FLIR PTU-E46-17P70T* and more finely tuned by the eyes' animation.

When the robot should look towards its human interaction partner, the application provides the 3D position of the human in world coordinates, which results from a multi-modal people tracker, a redesigned and extended version of [15]. The people tracker is integrating detections from laser-based leg detector and image-based people detectors working on the wide angle cameras.

Copying the human gaze, the head should not reflect every small change of the gaze angles. Rather should the head wait until one of the eye gaze angles' change exceeds a threshold and then fully reposition itself to the new gaze direction (for further details on people tracking with a PTU mounted camera see [16]).

In order to realize this behavior, the PTU is using its own controller which implements a hysteresis. The actual eye position then only has to consider the angular difference between the current head position and the gaze target (see Sect. 3.3).

4.3 Simulation of Mouth Movement

Since our application uses an active text-to-speech (TTS) system, the expectations of the interaction partner will be, that the voice goes along with some mouth movement. For this, the mouth LEDs can be used.

The audio stream of the TTS system is processed by a fast fourier algorithm which provides the frequency components of the audio signal to be played. From this power spectrum, a single frequency coefficient (around 440 Hz) is selected and given to the LED controller (see Sect. 3.4), which modulates the width of the mouth light bar accordingly.

5 Study on the Effect of Design Variants on the Perception of Help Requests

A web- and video-based questionnaire was used to compare variations of eye-expressions and LED-lighting for the situation of requesting help as a robot. As Dautenhahn et al. [17] showed, video-based evaluation of robots is feasible, as long as no physical interaction is involved.

5.1 Materials

Participants were shown video-snippets of the robot asking for help. A total of three variations for the eye-expressions were presented: a neutral, a sad, and a concentrated expression. This subset of expressions was chosen to match the needed states for the planned application of the robot, where it needs to express need for help, being busy and a neutral state. LED-lighting varied between seven levels of color and blink-frequency. The colors shown were blue and green, while the frequencies were 0 Hz, 0.5 Hz and 1 Hz. A control condition with all lights switched off was added as well. LED-lighting was designed as a between-groups measurement while eye-expression was measured within-subjects. After each video-snippet, participants rated the situation they experienced with several questionnaire items. Figure 7 shows a frame from one of the presented videos.

5.2 Measurement

The questionnaire consisted of several items. The first two questions asked the participants to answer, which expression they thought the robot showed and how strong the expression was. Nine alternatives were provided, of which eight were

Fig. 7. A frame with a sad eye-expression from one of the presented videos.

taken from the Facial Expression Identification [6,18]. The ninth alternative was added to include the concentrated expression as well. In addition, the participants could optionally mark one or more further expressions they thought could also be fitting. The second part of the questionnaire consisted of the four items of the hedonic subscale of the short version of the User Experience Questionnaire (S-UEQ) by Schrepp et al. [19]. The S-UEQ consists of a hedonic subscale, which focuses on the pleasantness of an interaction, and a pragmatic subscale, which focuses on subjective performance. The items for the pragmatic subscale were left out to shorten the overall length of the questionnaire, since there was no real interaction happening. The third part of the questionnaire contained two direct questions on the perceived politeness and rudeness of the robot's request as Salem et al. [20] proposed, and one item on how much time the participant would be willing to invest in this situation to help the robot after Pavey et al. [21]. The time had to be input with a unit-less slider anchored at little or much time.

5.3 Results

In the following sections, we first report the results and then interpret them separately. This serves to not mix objective results with subjective interpretation.

During four weeks, a total of 157 participants answered the questionnaire of whom 139 were included in the analysis. Exclusion was based on control items and for one participant on being too young.

Non-parametric tests had to be used, because of the violation of several assumptions for parametric tests. Regarding the initially planned analysis of covariance (ANCOVA), the assumption of homogeneity of regression slopes and the independence of independent variables from the covariates were violated. For the planned fallback analysis with several analyses of variance (ANOVA), severe deviations from homogeneity of variance and from normal distributions were found. Under these circumstances, the stability of type-I errors can not be guaranteed, resulting in inflated false-positives when reporting seemingly significant results. Thus, for the influence of the between-groups LED-lighting variation was analyzed with Kruskal-Wallis-Tests and post-hoc Mann-Whitney-U tests with Bonferroni-correction. The within-group eye-expressions were analyzed with Friedman's ANOVAs with post-hoc Wilcoxon signed-rank tests with Bonferroni-correction.

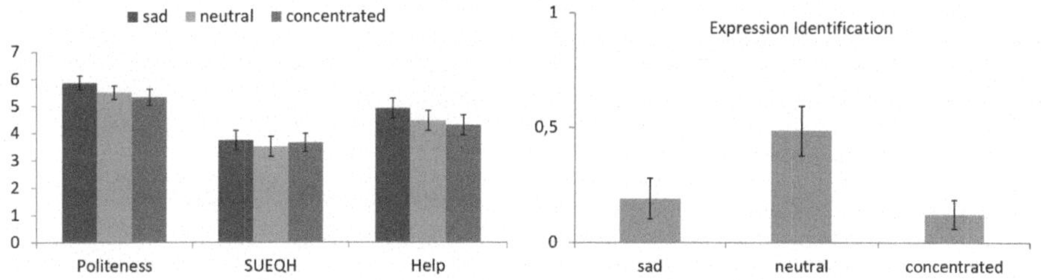

Fig. 8. *left*: Means and standard errors for the effect of eye expression on help intention, perceived politeness and the hedonic subscale of the S-UEQ. *right*: Means and standard errors for the effect of eye expression on expression identification, measured as the share of interpretations consistent with our intended expression.

LED-Lighting. A Kruskal-Wallis test with the independent variable LED-lighting yielded significant results for help intention with neutral eye-expression ($H(6) = 12.852$, $p = 0.045$) and concentrated eye-expression ($H(6) = 15.891$, $p = 0.014$). Post-hoc Mann-Whitney-U tests with LEDs off as a control group and with corrected significance niveau of $\alpha = 0.008\bar{3}$ were non-significant.

Eye-Expression. Friedman's ANOVAs for the three different eye-expressions yielded significant results for hedonic user experience $\chi^2(2) = 10.369$, $p = 0.006$, perceived politeness $\chi^2(2) = 42.792$, $p < 0.001$, facial expression identification $\chi^2(2) = 59.542$, $p < 0.001$ and help intention $\chi^2(2) = 53.043$, $p < 0.001$. Post-hoc Wilcoxon signed-rank tests used a corrected significance level of $\alpha = 0.01\bar{6}$.

For hedonic user experience, a significant difference was found between sad and neutral expression. Significant differences for perceived politeness were found between sad and neutral, as well as between concentrated and sad expressions. For the help intention, there were differences between all levels of eye-expressions. Means and standard errors for help intention, perceived politeness and hedonic UX are depicted in Fig. 8.

The expression identification showed significant differences between sad and neutral, and between concentrated and neutral. The test statistics for all significant post-hoc tests are shown in Table 3.

5.4 Discussion of Results

Due to violations of assumptions for parametric tests, non-parametric tests were used as an alternative. The a-priori sample size estimation based on parametric tests thus yielded a much smaller number of required participants than would have been needed for post-hoc tests to find present significant effects. Especially the significant results of the Kruskal-Wallis-H test for variations in LED-lighting could not be confirmed by post-hoc Mann-Whitney-U tests, most probably due to a lack of test power of the latter.

Table 3. Test statistics of significant post-hoc Wilcoxon signed-rank tests for pair-wise comparisons of the eye-expressions sad (S), neutral (N) and concentrated (C).

Dep. variable	Comparison	T	p	r
Help intention	S-N	771	<0.001	−0.319
	S-C	686	<0.001	−0.397
	N-C	1270	0.005	−0.167
Hedonic UX	S-N	1930.5	<0.001	−0.222
Perceived politeness	S-N	623.5	<0.001	−0.29
	S-C	514.5	<0.001	−0.337
Expression	S-N	867	<0.001	−0.331
Identification	N-C	318.5	<0.001	−0.417

The different eye-expressions differed in how well they were identified by the participants. The neutral expression scored higher in this regard than both the sad and concentrated expressions. This indicates optimization potential for the latter. Table 4 highlights how the expressions were interpreted by the participants instead. This knowledge will serve to further refine the respective expressions and to further distinguish them from the non-intended interpretations.

Table 4. Confusion matrix for the tested eye-expressions and answer options. All cells show the percentage of answers for the shown expression in the respective rows. The percentage of replies for the intended expression is underlined. Higher percentages of replies in non-intended categories than in the actually intended category are highlighted in gray.

Shown expression	Answer option								
	Angry	Happy	Sad	Fearful	Neutral	Surprised	Disgusted	Bored	Concentrated
Sad	1.80	2.88	<u>19.06</u>	32.37	32.73	3.24	0.36	0.72	6.82
Neutral	6.14	3.25	0.72	19.49	<u>48.74</u>	12.64	0.00	1.44	7.58
Concentrated	31.29	2.16	2.16	3.96	42.45	2.88	1.44	1.44	<u>12.23</u>

The hedonic UX was significantly higher for the sad eye-expression than for the neutral expression. The sad expression further increased the perceived politeness, compared to both the neutral and the concentrated expressions. Finally, help intention was highest for the sad eye-expression, followed by the neutral and the concentrated expressions. The results confirm and add to the positive effects of sad expressions on onlookers as reported by Marsh et al. [9].

6 Conclusion

In this paper, we have presented a new design and the technical realization of an expressive robot head, which comprises a PTU, 3D-printed shells, animated eyes displayed on two LCD-screens and three arrays with RGB-LEDs. The display solution and the LED matrix offer various possibilities for adaptation to the respective operating conditions and the interaction states to be conveyed. Therefore, it can easily be used in a conversation to communicate additional information to the dialog partner and to make the experience of interaction more pleasant.

The evaluation of an online study showed, that using the sad eye-expression yields benefits regarding the hedonic UX, perceived politeness, and help intention. Nonetheless, results regarding the expression identification performance suggest that there is further optimization potential for the sad as well as for the concentrated expression. Due to these findings, in future work we will develop more suitable eye animations and test the head in a complex application, where the robot has to navigate in a populated public building requesting help when it has to pass through closed doors or ride the elevator.

References

1. FRAME Project Website: Neuroinformatics and Cognitive Robotics Lab, Ilmenau University of Technology. http://www.frame-projekt.de. (in German)
2. Backhaus, N., Rosen, P.H., Scheidig, A., Gross, H.M., Wischniewski, S.: Help me, please?! Interaction design framework for needy mobile service robot. In: IEEE Workshop on Advanced Robotics and its Social Impacts (ARSO), pp. 54–61 (2018)
3. Liebner, J., Scheidig, A., Gross, H.M.: Now I need help! passing doors and using elevators as an assistance requiring robot. In: International Conference on Social Robotics (ICSR). Springer (2019)
4. Song, S., Yamada, S.: Expressing emotions through color, sound, and vibration with an appearance-constrained social robot. In: Proceedings of the 2017 ACM/IEEE International Conference on Human-Robot Interaction, pp. 2–11. ACM (2017)
5. Baraka, K., Veloso, M.M.: Mobile service robot state revealing through expressive lights: formalism, design, and evaluation. Int. J. Social Robot. **10**(1), 65–92 (2018)
6. Bennett, C.C., Šabanović, S.: Deriving minimal features for human-like facial expressions in robotic faces. Int. J. Social Robot. **6**(3), 367–381 (2014)
7. Ernest-Jones, M., Nettle, D., Bateson, M.: Effects of eye images on everyday cooperative behavior: a field experiment. Evol. Hum. Behav. **32**(3), 172–178 (2011)
8. Breazeal, C.: Emotion and sociable humanoid robots. Int. J. Hum Comput Stud. **59**(1–2), 119–155 (2003)
9. Marsh, A.A., Ambady, N., Kleck, R.E.: The effects of fear and anger facial expressions on approach-and avoidance-related behaviors. Emotion **5**(1), 119 (2005)
10. Lee, H.R., Šabanović, S., Stolterman, E.: How humanlike should a social robot be: a user-centered exploration. In: 2016 AAAI Spring Symposium Series (2016)
11. Mathur, M.B., Reichling, D.B.: Navigating a social world with robot partners: a quantitative cartography of the uncanny valley. Cognition **146**, 22–32 (2016)

12. Gross, H.-M., Boehme, H.-J., Schroeter, C., Mueller, S., Koenig, A., Martin, C., Merten, M., Bley, A.: ShopBot: progress in developing an interactive mobile shopping assistant for everyday use. In: IEEE International Conference on Systems, Man, and Cybernetics (SMC). IEEE, pp. 3471–3478 (2008)
13. Gross, H.-M., Mueller, S., Schroeter, C., Volkhardt, M., Scheidig, A., Debes, K., Richter, K., Doering, N.: Robot companion for domestic health assistance: Implementation, test and case study under everyday conditions in private apartments. In: IEEE/RSJ International Conference on Intelligent Robots and Systems (IROS), pp. 5992–5999. IEEE (2015)
14. ISO 60204–1:2016: Safety of machinery - Electrical equipment of machines - Part 1: General requirements. International Organization for Standardization, Geneva, CH, Standard (2016)
15. Volkhardt, M., Weinrich, Ch., Gross, H.-M.: Multi-modal people tracking on a mobile companion robot. In: European Conference on Mobile Robots (ECMR) (2013)
16. Vorndran, A., Trinh, T.Q., Mueller, S., Scheidig, A., Gross, H.M.: How to always keep an eye on the user with a mobile robot? In: International Symposium on Robotics (ISR), Munich, Germany. VDE Verlag, pp. 219–225 (2018)
17. Dautenhahn, K., Walters, M., Woods, S., Koay, K.L., Nehaniv, C.L., Sisbot, A., Alami, R., Siméon, T.: How may I serve you?: A robot companion approaching a seated person in a helping context. In: Proceedings of the 1st ACM SIGCHI/SIGART Conference on Human-robot Interaction. ACM, pp. 172–179 (2006)
18. Sosnowski, S., Bittermann, A., Kuhnlenz, K., Buss, M.: Design and evaluation of emotion-display EDDIE. In: 2006 IEEE/RSJ International Conference on Intelligent Robots and Systems, pp. 3113–3118. IEEE (2006)
19. Schrepp, M., Hinderks, A., Thomaschewski, J.: Design and evaluation of a short version of the user experience questionnaire (UEQ-S). IJIMAI **4**(6), 103–108 (2017)
20. Salem, M., Ziadee, M., Sakr, M.: Marhaba, how may I help you? Effects of politeness and culture on robot acceptance and anthropomorphization. In: 2014 9th ACM/IEEE International Conference on Human-Robot Interaction (HRI). IEEE, pp. 74–81 (2014)
21. Pavey, L., Greitemeyer, T., Sparks, P.: I help because I want to, not because you tell me to empathy increases autonomously motivated helping. Pers. Soc. Psychol. Bull. **38**(5), 681–689 (2012)

STEAM and Educational Robotics
Interdisciplinary Approaches to Robotics in Early Childhood and Primary Education

Lorenzo Manera(✉)

Department of Education and Human Sciences, University of Modena and Reggio Emilia, Viale Antonio Allegri, 9, 42121 Reggio Emilia, RE, Italy
lorenzo.manera@unimore.it

Abstract. In early childhood and elementary educational contexts, there is an increasing need to support the development of interdisciplinary skills such as creativity, digital literacy and critical thinking. Educational robotics can play a relevant role in this process, by providing contexts that involve design, construction and imagination. The STEAM (Science, Technology, Engineering, Art and Mathematics) approach offers a paradigm, which, by integrating educational robotics with scientific, digital and artistic perspectives, helps to undermine existing racial and gender inequality in STEM (Science, Technology, Engineering and Mathematics) and STEAM-related contexts. Some relevant experiences regarding the possibility to combine these fields in educational contexts show promising results. By discussing the most relevant difficulties related to the development of curricula where educational robotics finds a relevant place, this paper argues that the use of robotics as experimented with in the Reggio Emilia Approach allows to overstep most of the difficulties reported. Furthermore, it is argued that in order to be more effective, educational robotics might be considered by teachers not as a subject to teach by providing instructions but, instead, as a *medium* to allow children to explore the affordances of digital technologies in playful learning environments. Some practical examples of this perspective are reported and discussed in the final part of the paper.

Keywords: Steam · Educational robotics · Playful learning environments · Reggio Emilia Approach

1 New Paradigms in Educational Robotics

1.1 From STEM to STEAM

The STEAM (Science, Technology, Engineering, Art and Mathematics) approach offers interdisciplinary education involving fields such as science, technology, engineering, arts and math [1]. Preparing children to live in an increasingly technological and global society requires literacy processes that empower young students to collaborate effectively in teams across different disciplines. In recent years, particularly in Europe, several documents have been published in this regard. In particular, they indicate the need to support the development of interdisciplinary skills such as creativity, digital literacy and critical

F. Ferraguti et al. (Eds.): HFR 2019, SPAR 12, pp. 103–109, 2020.
https://doi.org/10.1007/978-3-030-42026-0_8

thinking [2]. If we consider the policies developed in Europe, we can see that STEM (Science, Technology, Engineering and Mathematics) education has been interpreted mainly as a useful approach to ensure the development of cognitive competences and skills [3]. Despite the relevance of these guidelines, the risk of excluding human sciences in the process of defining development strategies in the European context needs to be taken into account [4]. In fact, the emergence of instances related to recognizing the role that critical thinking and creativity can play in developing strategic skills has led to the emergence of the STEAM education paradigm, understood as a vast set of educational and design strategies. These strategies mainly aim to promote a dialogue between scientific disciplines and the related interweaving of skills in a global perspective of development [5]. The transition from STEM to STEAM has been interpreted as an innovative process, aimed at developing a field of research that includes the presence of scientific and humanistic cultures [6] with a view to overall cross-curricular innovation to support the processes of meaning-making in the different educational contexts involved [7]. The scientific literature underlines the existence of racial and gender inequality in STEM and STEAM related contexts? [8]. It is therefore necessary to provide broader possibilities for both boys and girls to start engaging in STEAM-related activities starting from early childhood education contexts. In this regard, Educational Robotics can play a relevant role, by providing contexts that involve design, construction, creativity, imagination and critical thinking [9].

1.2 Educational Robotics in STEAM Educational Contexts

Progress in robotics development has created new opportunities and challenges for early childhood educational settings. As underlined in scientific literature [10], from a pedagogical point of view, Educational Robotics refers to the epistemological paradigm of constructivism and generally involves both programming and construction phases. Educational Robotics (ER) refers, in particular, to a convergence between robotics and pedagogy aimed at creating meaningful robotics experiences from early childhood.

Coherently with this paradigm, these phases should generally be proposed in a playful environment, allowing children to explore and develop projects and ideas. This process can help children develop their creativity, by connecting their knowledge in different fields [11] as they start to deal with abstract concepts such as time, movement and space. Offering learning contexts where children have the possibility to take their time to develop ideas and attempt to create new projects also fosters relevant metacognitive processes, thanks to the possibility to re-think the different steps and verify which ones worked [12].

1.3 Educational Robotics, Aesthetic Elements and Play

The broad availability of basic computational applications, such as the Arduino microcontroller board LiliPad, has helped children to connect and study aesthetic elements, programming and technology, as scientific literature shows [13]. These applications have also encouraged young girls to get more involved in science and technology [14]. Diverse forms of artistic expression have been successfully combined with robotics.

Wearable computing devices [15], art-robotics [16] and interdisciplinary courses combining art, programming and engineering [11] have shown potential in engaging children and students. Furthermore, Educational Robotics combines play with learning and thus learning becomes a fun activity that is easier, more interesting and more efficient [12]. The elements outlined above, in particular the connection between artistic and scientific perspectives and the playful environments where children enact their learning processes, can help to define innovative learning environments where the importance of providing playful and engaging contexts, especially for young children, is underlined. In fact, when exploring digital and robotic technologies, play appears to be a fundamental element. Thanks to this attitude, children actively construct a model of the world. The process of understanding robotics can happen in a reciprocal, shared way, in a dimension characterized by a harmonious balance between learning and teaching. By creating new meanings in playful processes that involve digital technologies, children give life to authentic cognitive processes in relation to learning contexts. In the age of neo-technologies, digital media constitutes one of the main experiential environments, as the etymological root *medius* indicates: a place for relationships, exchange, and therefore places for play and learning that characterize our era. Learning to actively and critically engage with digital technologies and robotics can become, as Seymour Papert suggested, a natural process for children [17]. Symbolically, imagination and intellect freely interact thanks to experiences and perceptions, giving form to new ideas in both adults and children, creating innovative, relational properties between the environment and the technologies that help generate meaning.

1.4 Educational Technologies in Constructivist Learning Environments

If offered the possibility to freely engage with digital technologies, children explore the properties and different relational dimensions enabled by the contexts offered. This activates cognitive and imaginative faculties, leading to something else with respect to the contextual dimension of the experience and the perception of it, provoking new ideas and possibilities linked to the construction of meaning that characterize playful learning experiences. Like play, "learning processes are broad, interconnected and dynamic" [18]. The Reggio Emilia Approach to education has been proposing educational robotics in playful environments since the 1980's [19], in particular thanks to the inspiration and collaboration established with Seymour Papert. The scientific literature reporting successful Educational Robotics experiences within STEM education and the Reggio Emilia Approach is based on the pioneering work of Seymour Papert [20], who advocated the use of computers and robots to support the processes of knowledge building, creating new ideas and artifacts and sharing them.

By developing an educational constructivist theory, Papert introduced "Logo", a basic coding language, aimed at enhancing active learning and offering hands-on activities. Logo was the first easily accessible coding language explicitly designed for learning. It allowed children to develop sophisticated ideas about programming [21] and to explore projects and ideas.

Afterwards, in 1970's, the MIT Artificial Intelligence's research group developed the first Turtle (R). Papert added a geometrical perspective that reflected the robot's perspective on space to Logo, called Turtle Graphics (R) [22]. In 1984, a research project was developed in the Municipal Preschools of Reggio Emilia dedicated to exploring Turtle [23]. In her reflections on the research project, the *Pedagogista* Paola Cagliari wrote that most relevant aspect of the experience was the fact that the children learned a new language that allowed them to develop a "dialogue with the environment" [24]. In the same year, the CEO of the LEGO company, Kjeld Kirk Kristiansen, saw a video interview with Papert. By showing how children could use Logo, he explained his pedagogical approach. This led Kjeld Kirk Kristiansen to meet Papert and then fund a research project that resulted in the official presentation of Lego Mindstorms (R) in 1998 [25]. In the early 2000s, three Municipal Preschools in Reggio Emilia took part in a research project dedicated to testing Lego Mindstorms (R) in early childhood learning environments [26]. During the research, the children created a large map of an imaginary city, and then programmed the robots to move to different parts of it. During the second phase of the project, a group of children tried to understand if it was possible to program a robot in order to make it bring some bread to feed some little birds. These examples suggest the possibility of proposing robotics not as a discipline to be taught by providing instructions but rather as a *medium* that allowed children to face challenges and problems that they found relevant. In the Reggio Emilia Approach, in fact, the cognitive act of play can be understood as "a creative act that involves the assumption of responsibility, autonomy and freedom", in a dimension "of solidarity between play and learning" [25], which can be referred to a playful learning. The encounter with Educational Robotics, therefore, should be proposed as an intrinsically motivated act, which involves "design thinking" and where greater importance is attributed to the process rather than the result. Thus, the idea of children exploring Educational Robotics becomes an occasion for creating, negotiating and understanding meanings that change and are open to children's imaginations. Based on these ideas, resonance with the Learning Through Play (LtP) paradigm, developed by the Lego Foundation, is evident. According to LtP, optimal learning happens when experiences are joyful, help children find meaning in what they are doing, involve active, engaged, hands-on thinking, involve iterative thinking (e.g., experimentation, hypothesis testing) and involve social interaction. These elements also define the educational experience that has characterized the Reggio Emilia Approach, underlining the value of play, recognized as an essential element of any authentic cognitive and educational process. In this process, the dimension of time is essential: it allows for consolidating qualities of play, such as convergent and divergent thinking, without excluding aesthetic elements.

By proposing robotics in complex learning environments, the possibilities for children to find new affordances and connections between different elements of the environments are multiplied. This process shifts children's attention from the interface to more complex situations, where they can solve actual problems or act on several representational levels by trying, modifying and making mistakes.

The possibilities offered by the use of robotics in learning environments can help teachers and children in both problem posing and problem solving, by reformulating situations, embedding new elements and trying different solutions.

For example, exploring the new version of Scratch, the open platform developed by the Lifelong Kindergarten group (MIT Media Lab), whose latest version can used to program robots such as Lego Mindstorms, children can apply Papert's procedural thinking by executing different sequences at the same time, in order to explore connections between different elements.

By embedding programming language, these tools can promote the modelling of situations, the resolution of problems and the potential development of problems, fostering for investigation and research skills.

1.5 Conclusions

Despite great changes in early childhood educational paradigms toward constructivist epistemology that have evolved since Papert's first educational robots and ideas were developed, robotics has not still found a relevant place in pre- and primary school curricula [27]. The largest difficulties reported in scientific literature are related to different elements [28]. Firstly, the fact that these robotic activities require constructivist methodologies, allowing flexibility in curricula and openness to novelty. Moreover, a strong gender bias informs educational robotics. Finally, there is a lack of curricula involving critical thinking, teamwork and creativity. In fact, Educational Robotics is often an activity aimed to let children build a robot step-by-step, following guidance provided by the teacher, rather than as a medium that allows a greater student engagement in developing projects and ideas. In order to address the problems mentioned, the Reggio Emilia Approach offers several opportunities. In a recent research project conducted in Reggio-inspired weekly after-school program [29], a group of children developed the idea of building a garbage detector robot, similar to the garbage truck they used to seeing in their everyday lives, because they thought they needed some help cleaning their space.

This project, among others, offers an example of how the Reggio Emilia Approach can inform constructivist use of robotics in different ways: (1) by offering an environment that supports children's authentic learning, (2) by involving children in the development of an emerging curriculum, (3) by offering the possibility to develop ideas in meaningful situations and (4) by sustaining the development of multimodal competences in children.

This educational approach moves from a context where robotics is "taught" to children to a paradigm where the possibilities offered by Educational Robotics are explored. For instance, the Reggio Approach provides innovative learning environments, where children are offered settings where they can explore different ideas and make hypotheses, therefore representing an ideal context where robotics can be intermixed with different disciplines, as in a STEAM learning environment. In such learning environments, children can explore, among other materials, Lego Mindstorms sensors and create connections with artistic elements. This offers a possible solution for the gender bias that affects Educational Robotics. Furthermore, Reggio Emilia Preschools offer situations where children can create paths by using different types of materials and then add a color sensor so that the robot can explore the perceptual properties that it recognizes. By proving these opportunities, the possibilities offered by robots become part of the affordances of a learning environment, and therefore help to support children's learning.

Furthermore, children can explore the aesthetic qualities of robotic components and combine them with unstructured materials in immersive environments. With a light sensor, children can create basic systems to move small plants to bright areas. What is relevant in these kinds of contexts is that learning to code is not the goal of the activity:

it becomes a medium that allows children to explore their curiosity and create different projects. What emerges is the importance of the sense-making process, fostered by creating learning environments where robotics can be perceived by teachers and children not as a discipline to learn, but as a tool able to sustain interdisciplinary investigations.

References

1. Babaci-Wilhite, Z.: Promoting Language and STEAM as Human Rights in Education. Springer, Singapore (2019)
2. European Commission: Science education for responsible citizenship: report to the European Commission of the expert group on science education. Directorate-General for Research and Innovation science with and for society: EUR 26893 EN (2015)
3. European Commission: Does Europe need more STEM graduates. Final report. Danish Technological Institute, Danish, pp. 1–68 (2015)
4. Van den Doel, H.W.: The future of the Social Sciences and Humanities in Europe: collected LERU papers on the SSH research agenda. League of European Research Universities (LERU) (2013)
5. De la Garza, A., Travis, C. (eds.): The STEAM Revolution: Transdisciplinary Approaches to Science, Technology, Engineering, Arts, Humanities and Mathematics. Springer, Cham (2019)
6. Henriksen, D.: Creating STEAM with design thinking: beyond STEM and arts integration. STEAM J. **3**(1), 11 (2017)
7. Land, M.H.: Full STEAM ahead: the benefits of integrating the arts into STEM. Procedia Comput. Sci. **20**, 547–552 (2013)
8. Riegle-Crumb, C., King, B., Irizarry, Y.: Does STEM stand out? Examining racial/ethnic gaps in persistence across postsecondary fields. Educ. Researcher **48**(3), 133–144 (2019)
9. Charoula, A., Valanides, N.: Developing young children's computational thinking with educational robotics: an interaction effect between gender and scaffolding strategy. Comput. Hum. Behav. 1–60 (2019)
10. Alimisis, D., Kynigos, C.: Constructionism and robotics in education. In: Alimisis, D. (ed.) Teacher Education on Robotics-Enhanced Constructivist Pedagogical Methods, Athens, Greece (2009)
11. Brunvand, E., Stout, P.: Kinetic art and embedded systems: a natural collaboration. In: Proceedings of the 42nd ACM Technical Symposium on Computer Science Education, pp. 323–328. ACM (2011)
12. Stergiopoulou, M., Karatrantou, A., Panagiotakopoulos, C.: Educational robotics and STEM education in primary education: a pilot study using the H&S electronic systems platform. In: International Conference EduRobotics, pp. 88–103. Springer, Cham (2016)
13. Buechley, L., Eisenberg, M., Catchen, J., Crockett, A.: The LilyPad Arduino: using computational textiles to investigate engagement, aesthetics, and diversity in computer science education. In: Proceedings of the SIGCHI Conference on Human Factors in Computing Systems, pp. 423–432. ACM (2008)
14. Montero, C.S., Jormanainen, I.: Theater meets robot–toward inclusive STEAM education. In: International Conference EduRobotics 2016, pp. 34–40. Springer, Cham (2016)
15. Lau, W.Y., Ngai, G., Chan, S.C.F., Cheung, J.C.Y.: Learning programming through fashion and design: a pilot summer course in wearable computing for middle school students. In: Proceedings of the 40th ACM Technical Symposium on Computer Science Education (SIGCSE 2009). ACM, New York (2009)

16. Kim, J.H., Coluntino, D., Martin, F.G., Silka, L., Yanco, H.A.: Artbotics: community-based collaborative art and technology education. In: ACM SIGGRAPH 2007 Educators Program (SIGGRAPH 2007). ACM, New York (2007)
17. Papert, S.: Mindstorms: Children, Computers, and Powerful Ideas. Basic Books Inc., New York (1980)
18. Zosh, J.N., Emily, J., Jensen, H., Liu, C., Neale, D., Hirsh-Pasek, K., Lynneth Solis, S., Whitebread, D.: Learning through play: a review of the evidence. LEGO Foundation (2017)
19. Cagliari, P.: Notes on research, in border-crossing, Reggio Emilia, Reggio Children srl, pp. 10–13 (2019)
20. Scaradozzi, D., Screpanti, L., Cesaretti, L.: Towards a definition of educational robotics: a classification of tools, experiences and assessments. In: Daniela, L. (ed.) Smart Learning with Educational Robotics, pp. 63–92. Springer, Cham (2019)
21. Papert, S.: Talking turtle. Horizon, Part 1–02:48 to 03:10. BBC and the Open University (1983)
22. Ableson, H., diSessa, A.A.: Turtle Geometry. MIT Press, Cambridge (1981)
23. Cagliari, P.: Notes on research, in border-crossing, Reggio Emilia, Reggio Children srl, p. 11 (2019)
24. Immovilli, G.: Quando due intelligenze si incontrano. Bambini, pp. 76–80 (1985)
25. http://hackeducation.com/2015/04/10/mindstorms
26. Giacopini, E.: Quando due intelligenze si incontrano – parte III, in "Bambini", vol. 12, pp. 82–87 (1986)
27. Rinaldi, C.: In Dialogue with Reggio Emilia. Routledge, New York (2006)
28. Branscombe, N.A., Burcham, J.G., Castle, K., Surbeck, E.: Early Childhood Curriculum: A Constructivist Perspective. Routledge, New York (2013)
29. Alimisis, D., Moro, M.: Special issue on educational robotics. Robot. Auton. Syst. **77**(C), 74–75 (2016)
30. Meacham, S., Atwood-Blaine, D.: Early childhood robotics: a Lego robotics club inspired by Reggio Emilia supports children's authentic learning. Sci. Child. **56**, 57–61 (2018)

Grasp-Oriented Myoelectric Interfaces for Robotic Hands: A Minimal-Training Synergy-Based Framework for Intent Detection, Control and Perception

Roberto Meattini[1(✉)], Luigi Biagiotti[2], Gianluca Palli[1], and Claudio Melchiorri[1]

[1] Department of Electrical, Electronic, and Information Engineering "Guglielmo Marconi", University of Bologna, Viale del Risorgimento 3, 40136 Bologna, Italy
roberto.meattini2@unibo.it

[2] Department of Engineering "Enzo Ferrari", University of Modena and Reggio Emilia, Strada Vignolese 905, 41125 Modena, Italy
luigi.biagiotti@unimore.it

Abstract. In this paper, a general framework for myoelectric human-robot interfaces for grasp-oriented tasks is presented. In particular, the aspects of intent detection, control and perception are all three integrated in the framework. Particular attention is devoted to provide a minimal-training procedure for the user, and to convey sensory feedback information from contact force sensors in a compact manner, by means of a single vibration motor to foster practical real-world usage of the interface. Specifically, in the proposed framework, the user's motor intention is estimated from electromyographic measurements exploiting the concept of muscular synergy, together with an unsupervised learning approach that enables a short and simple algorithm calibration. Then, the postural synergy concept is exploited in an encoding-decoding fashion, in order to communicate the user's hand closure level to the robotic hand, i.e. exploiting kinematic dimensionality reduction. Moreover, a rough estimation of the internal grasping force applied by the robotic hand is exploited for the generation of a sensory substitution feedback, which is applied to the user by means of vibrotactile stimulation. After describing the general structure of the proposed framework, we show a practical design for a real interface implementation. Finally, experimental tests with 4 naïve subjects are reported demonstrating the actual effectiveness of the approach – in terms of mean and standard deviation of grasp strength reference trackings – in fine object grasping tasks.

Keywords: Robotic hand · Grasping · Electromyography · Postural synergies · Vibrotactile feedback

F. Ferraguti et al. (Eds.): HFR 2019, SPAR 12, pp. 110–124, 2020.
https://doi.org/10.1007/978-3-030-42026-0_9

1 Introduction and Research Context

Robotics is currently experiencing a growing demand for a closer cooperation, interaction and integration with humans. Driven by this need, in the recent years a new generation of robots characterized by natural and intuitive human-machine interfaces was developed, in which the design steered towards human-inspired solutions, as a road to replicate the human ability and flexibility in performing motor tasks. A key element in the general framework of Human-Robot Interfaces (HRIs) is the so-called *human-in-the-loop* integration, where the human-machine interaction is a fundamental and constitutive element of the system design.

To enable a natural interaction between humans and robots, a major challenge is to leverage on signals acquired from the human body to communicate with the machines. Surface skin ElectroMyoGraphic (sEMG) signals are widely popular in several scenarios, ranging from rehabilitation to robot control, and in particular they are often exploited as an interface between human subjects and robotic hands. With this respect, it is worth noticing that the human hand shows remarkable capabilities in object manipulation, whereas, in contrast, robotic devices are characterized by poor performance in fine physical interaction with objects and their control is still a big challenge in the scientific community [1]. However, the configuration space dimension of the human hand can be reduced – considering a set of common grasping tasks – thanks to the concept of *postural synergies* [2], according to which the grasp synthesis for a large set of daily-life objects is achievable by a linear combination of a limited set of dominant hand postures, commonly referred as postural synergies or *eigenpostures*.

Aiming to reproduce the human's intention on a robotic device, machine learning techniques are widely used to improve the correlation between muscle activity measured through sEMG on a human subject and the corresponding motion commanded to a robotic hand. Several authors explore different classifiers, such as neural networks [3], Linear Discriminant Analysis (LDA) [4] or Support Vector Machines (SVMs) [5]. However, classifier-based approaches present several drawbacks, such as long training periods, large demand of human time for data labeling and limited set of implementable task or action classes, resulting in limited affordability and flexibility of the human-machine interface.

To overcome these limitations, several supervised machine-learning-based regression algorithms have been implemented by the research community. The advantage of this method lies in the possibility of continuously controlling the DoF, so that ambiguities in input signals do not lead to totally unwanted results as in case of classifiers. However, despite several positive outcomes, regression-based ME HRIs presents several reliability issues. This is likely due to two main assumptions on which this method relies: invariance of the sEMG signals with respect to subjects and time [6] and accuracy/reliability of the data labeling for the algorithm training. These conditions are hardly satisfied in practice – both for intact users and rehabilitation applications – resulting in degraded performance and instability in the robotic hand interface. Additionally, the knowledge gained in the past decades about the neurophysiological processes underlying natural movements, such as muscular synergies, is not exploited in the

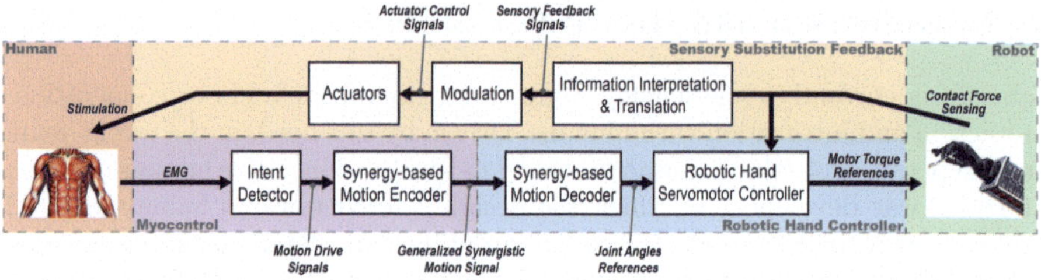

Fig. 1. The general framework for grasp-oriented myoelectric HRIs presented in this article.

supervised machine learning framework. For these reasons, in the present work we propose an ME interface for robotic hands based on unsupervised machine learning technique in which data regression is implemented by means of a matrix factorization approach, thanks to the modelling of the human's neuro-motor system combined with muscular and postural synergy concepts. In this way, the training procedure and the data labeling is not necessary, and only an extremely simple and fast calibration data acquisition is required.

Another important aspect related to the use of myocontrolled robotic hands for the execution of grasping tasks is the availability of an artificial sensory feedback related to the grasp strength. The most common methods are patterned electrotactile, mechanotactile, or vibrotactile stimulation of the skin, in accordance with the general approach known as *sensory substitution* [7]. In the present work, we focus on the vibrotactile feedback, i.e. the use of vibration motors to deliver skin mechanical stimulations. Note that vibrotactile feedback is not new at all, since it has been already studied in several works [8–10]. However, despite the fact that many works answered the question *if* vibrotactile force feedback can help the user in executing grasping operations (and the problem is still open [11]), in this work – since a grasp-oriented interface for practical applications is concerned – we will focus on *how to* enable the user to precisely reproduce grasp strength reference values *exploiting vibrotactile feedback* with a myocontrolled robotic hand. For this reason, the main contribution of this work consists in the presentation of a complete myoelectric human-robot interface framework for robotic hands control, in which the user provides commands to the robotic device, equipped with contact force sensors mounted on the fingertips, and receives an information that synthesizes the interaction force during grasping via a simple vibrotactile device. In this framework, the detection, the control and the perception aspects are all considered together with the design and implementation issues in a synergy-based approach. Other important features we embedded in the proposed framework are the minimal training effort, the independent force regulation of the fingers to adaptation the grasp to objects, a sensory substitution approach based on contact force sensors measurements for the use with EMG signals.

In the following, the proposed framework is presented together with its practical implementation. Moreover, the experimental evaluation of the framework is carried out and the results are reported and analyzed.

2 Synergy-Based Framework for Myoelectric HRI

2.1 General Structure Overview

The framework presented in this work for grasp-oriented myoelectric interfaces is depicted, in its structure, in Fig. 1. In the figure, the different blocks composing the bidirectional chain of the human-robot's interaction are represented, subdivided in the areas of (*i*) Human, (*ii*) Myocontrol, (*iii*) Robotic Hand Controller, (*iv*) Robot and (*v*) Sensory Substitution Feedback.

The *Human* area is the interface zone in which the neuro-motor information is gathered from the user by means of measurements related to some physiological activity – in particular, in this framework we specifically consider the sEMG signal. This general concept is discussed from a design point of view and implementation case in Subsect. 2.2, in terms of electrodes, data acquisition and signal processing. At the same time, also the interfacing to the sensory feedback stimulation is here located, as reported for the real case scenario in Subsect. 2.4. In the *Myocontrol* area, the human's intent interpretation takes place: in the proposed framework this is based on a modelling of the sEMG-wise intent generation, lying on the exploitation of the muscular synergy and neural drive concepts, with the specific aim of applying unsupervised learning techniques in order to have minimal training sessions of the myocontrol algorithm by the user of the interface. Here, also the concept of postural synergy is exploited, in order to encode the human kinematic information in a neuro-physiologically compact form (i.e.: dimensionality reduction). Accordingly, the *Robotic Hand Controller* area is composed by a synergy-based decoder of the motion, that in this way can be provided as input for the lower-level servomotor controller of the anthropomorphic grasping device, providing appropriate torques according to the requested behaviour – details of this concepts in a practical sense are illustrated in Subsect. 2.3. The *Robot* area includes the robotic hand, which is equipped with contact force sensors both for sensory feedback generation and for servomotor controller in order to adapt to different object shapes (hardware and design aspects are reported in Subsect. 2.3). Then, the interaction forces exchanged between robot and external world are interpreted and modulated within the *Sensory Feedback Substitution* area, in which the signals detected by the contact sensors are translated into a meaningful stimulation for the human, which is provided by a vibrotactile actuator for the feedback delivery (the practical implementation is reported in Subsect. 2.4).

2.2 Factorization-Based Myocontrol

SEMG Acquisition Setup and Pre-processing. The sEMG signal acquisition hardware is based on the Cerebro board [12]. The data are acquired at 1 kHz

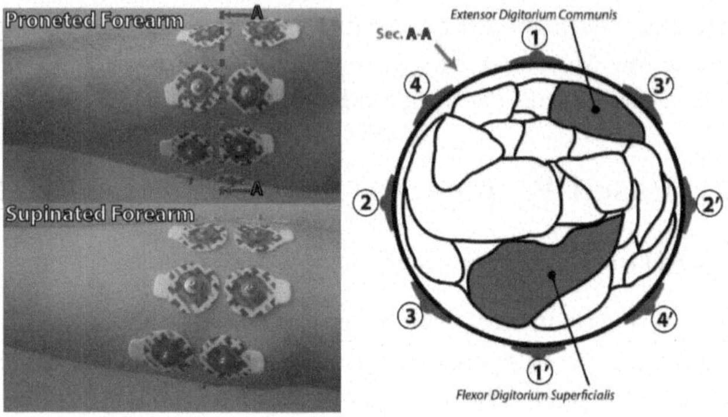

Fig. 2. Location of the sEMG electrodes on the forearm.

and streamed to a nearby computer using the onboard Bluetooth interface. The placement of the electrodes is conceived to cover the portion around the forearm in proximity of the *Flexor Digitorum Superficialis* and the *Extensor Digitorum Communis* muscle bellies. As can be seen in Fig. 2, 16 disposable surface skin electrodes are arranged to form an armband composed by 8 pairs of differential electrodes uniformly distributed, obtaining in this way 8 sEMG signals. Then, a standard online filtering procedure is implemented, consisting of: *(i)* a 50 Hz notch filter; *(ii)* a 20 Hz high-pass filter; and finally *(iii)* the root mean square (RMS) value of the signal is calculated over a 200 ms window.

Synergy-Based Model of Human's Hand Closure Motion Control. Let us consider the RMS value of the online 8-channel sEMG acquisition (see Sect. 2.2) $E_{\text{on}}(t) \in \mathbb{R}^{8\times 1}$. We suppose that at each time instant this multidimensional signal can be expressed as the product of the *muscular synergy matrix* $S_M \in \mathbb{R}^{8\times 2d}$ and the neural drives $U_{\text{on}}(t) \in \mathbb{R}^{2d\times 1}$ [6], where d is the number of DoF activated by the muscles that generate the muscle activity $E_{\text{on}}(t)$:

$$E_{\text{on}}(t) = S_M U_{\text{on}}(t). \tag{1}$$

Note that, while $E_{\text{on}}(t)$ is known, in Eq. (1) S_M and $U_{\text{on}}(t)$ are the unknowns.

The concept of postural synergies is exploited in order to consider the overall opening and closing of the human hand as a unique DoF, in particular the only one activated by the muscles generating the muscle activity $E_{\text{on}}(t)$. Moreover, such movement can be considered as a single synergistic motion, given by one specific postural synergy. Therefore, according to this consideration, we assume $E_{\text{on}}(t)$ expresses the muscle activity during the regulation of such single synergistic DoF, which means that $d = 1$. With these assumptions in mind, it is possible to consider two antagonistic actions, namely "generalized" flexion and extension, for the actuation of that synergistic DoF, i.e. the hand closure. Similarly, we can also consider two neural drives that activate such antagonistic actions. In this way, exploiting the antagonistic actuation concept [13], it is possible to write the matrices in Eq. (1) as

$$S_M = \begin{bmatrix} s_e \; s_f \end{bmatrix}, \quad U_{\text{on}}(t) = \begin{bmatrix} u_e(t) \\ u_f(t) \end{bmatrix}, \tag{2}$$

where s_e, $s_f \in \mathbb{R}^{8\times 1}$ are the extension and flexion components of the muscular synergy matrix and $u_e(t)$, $u_f(t) \in \mathbb{R}$ are the extension and flexion components of the neural drives. Equations (2) and (1) represent the sEMG model during the synergistic opening/closing motion of the human hand. This model can be exploited *online* to detect the user's intent to open/close the hand. To this end, it is necessary to estimate the muscular synergy matrix S_M from a sEMG calibration recording (filtered as explained in Sect. 2.2) containing n data samples collected in a matrix $E_{\text{off}} \in \mathbb{R}^{8\times n}$, corresponding to the muscle activity during the execution of a simple open and close motion of the hand for two times. Therefore, on the basis of (2) and (1), E_{off} can be expressed as

$$E_{\text{off}} = S_M U_{\text{off}}, \tag{3}$$

with

$$S_M = \begin{bmatrix} s_e \; s_f \end{bmatrix}, \quad U_{\text{off}} = \begin{bmatrix} u_e^{\text{T}} \\ u_f^{\text{T}} \end{bmatrix}, \tag{4}$$

where $U_{\text{off}} \in \mathbb{R}^{2\times n}$ is the matrix of the offline neural drives composed of the extension and flexion components u_e, $u_f \in \mathbb{R}^{n\times 1}$ (note that they are not a function of time in this case). Taking into account (3), and on the light of the *a priori* knowledge of a unique synergistic DoF activated during the calibration process, it is possible to solve (4) with respect to S_M and U_{off}. Consequently, a factorization algorithm can be applied to compute the S_M. In this way, once the muscular synergy matrix is estimated, it is possible to use its pseudo-inverse S_M^+ to online compute the neural drives of the human hand opening/closing motion as

$$U_{\text{on}}(t) = S_M^+ E_{\text{on}}(t).$$

In order to obtain a proportional control signals for the grasp control, the concept of the antagonistic actuation model [13] is exploited. Two antagonistic actions are considered, representing two groups of flexor and extensor hand muscles located in the forearm, that control the synergistic DoF and the stiffness level of the hand. According to this concept, the online estimated neural drives are linearly combined to obtain the closure hand reference $\rho^{\text{ref}}(t)$ control signals as

$$\rho^{\text{ref}}(t) = a\,(u_e(t) - u_f(t) + b) \tag{5}$$

where a and b are proper constant values that scale ρ^{ref} in the range $[-0.5, 0.5]$.

2.3 Human-Like Robotic Hand Controller

The UB Hand: Structure and Sensory Equipment. The University of Bologna Hand (UB Hand) is a robotic hand with an anthropomorphic structure [14,15], see Fig. 3(a). Each one of the 5 fingers, characterized by 3 DoF, is fully actuated via 5 servomotors (Dynamixel RX-24F) located in the forearm whose motion is transmitted by inelastic tendons, which are properly coupled and routed in the finger structure. Furthermore, the UB Hand is equipped with three semi-spherical contact force sensors (OMD-20-SE-40N Optoforce [16]), able to detect the magnitude of forces acting along the x, y and z directions as shown in Fig. 3(b), namely f_x, f_y, f_z. The three sensors are mounted on the fingertip of the thumb, index and middle fingers by means of a 3D-printed support (see Fig. 3). In this work, the norm of the force vectors measured the sensors is used. They are denoted as $|f_i| = \sqrt{f_{x,i}^2 + f_{y,i}^2 + f_{z,i}^2}$, where the subscript $i = \{T, I, M\}$ refers to Thumb, Index and Middle fingers, respectively.

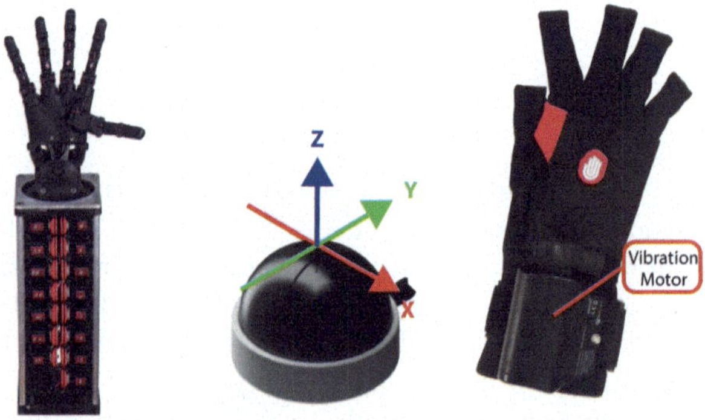

(a) The UB Hand. (b) Force sensor. (c) Noitom Hi5 Glove.

Fig. 3. The UB Hand, contact force sensor and Hi5 Glove.

Synergy-Based Hierarchical Controller. In order to control the overall robotic hand, a proper mapping between the n_m tendons actuating the joints, each one driven by a servomotor, and the n_J joints of all the hand fingers exists. This control architecture is characterized by a three-level hierarchical structure, with three different levels: motor, joint and synergy. At the lowest level, a stiffness control has been implemented [17]. Therefore, the Dynamixel servomotors are actuated according to

$$\tau_m(t) = K_m(\theta_m^{\mathrm{ref}}(t) - \theta_m(t)) \ , \tag{6}$$

where $\tau_m \in \mathbb{R}^{n_m}$ is the vector of motor torques, $K_m \in \mathbb{R}^{n_m \times n_m}$ is the diagonal matrix defining the motor stiffness, and θ_m^{ref}, $\theta_m \in \mathbb{R}^{n_m}$ are the motors reference and the actual position, respectively. At the joint level, the mechanical configuration of the tendon-based transmission system defines a linear mapping $H \in \mathbb{R}^{n_m \times n_J}$ between joint and motor configuration variables (for more details refer to [14,15]). This mapping can be used to deduce the reference angle of the motors corresponding to a given joint configuration θ_J^{ref}, i.e.

$$\theta_m^{\text{ref}}(t) = H\ \theta_J^{\text{ref}}(t). \tag{7}$$

By combining (6) and (7) it is possible to deduce the stiffness relation induced at the joint level by the motor control:

$$\tau_J(t) = K_J(\theta_J^{\text{ref}}(t) - \theta_J(t))\ , \tag{8}$$

where $K_J = H^T K_m H \in \mathbb{R}^{n_J \times n_J}$ defines the joint stiffness matrix. Finally, at the the higher synergy level, the joint reference configuration θ_J^{ref} is computed as reported in the following.

Let

$$\dot{\theta}_J^{\text{ref}} = \left[\dot{\theta}_T^{\text{ref}}\ \dot{\theta}_I^{\text{ref}}\ \dot{\theta}_M^{\text{ref}}\ \dot{\theta}_R^{\text{ref}}\ \dot{\theta}_P^{\text{ref}}\right]^{\text{T}} \in \mathbb{R}^{n_J}$$

be the vector of the robot hand joint velocities, where $\dot{\theta}_i^{\text{ref}} = \left[\dot{\theta}_{i,1}^{\text{ref}}\ \dot{\theta}_{i,2}^{\text{ref}}\ \dot{\theta}_{i,3}^{\text{ref}}\right]^{\text{T}}$ denotes the velocities of a single finger, with $i = \{T, I, M, R, P\}$ indicating thumb, index, middle, ring and pinkie fingers, respectively. The instantaneous joint reference configuration $\theta_J^{\text{ref}}(t)$, defined according to the concept of postural synergies, is computed as

$$\dot{\theta}_J^{\text{ref}}(t) = K_{\text{gain}}\ S_P\ \alpha_v\ \sigma(t). \tag{9}$$

The input variable $\sigma(t)$, that appears in (9), is the myocontrol-driven signal obtained by applying a dead-zone to $\rho^{\text{ref}}(t)$ defined in (5), that is

$$\sigma(t) = \begin{cases} \rho^{\text{ref}}(t) - \bar{\rho}_1, & \text{if } \rho^{\text{ref}}(t) \leq \bar{\rho}_1 \\ 0, & \text{if } \bar{\rho}_1 < \rho^{\text{ref}}(t) < \bar{\rho}_2\ , \\ \rho^{\text{ref}}(t) - \bar{\rho}_2, & \text{if } \rho^{\text{ref}}(t) \geq \bar{\rho}_2 \end{cases} \tag{10}$$

where $\bar{\rho}_1$ and $\bar{\rho}_2$ are two proper constant threshold values with $\bar{\rho}_2 > \bar{\rho}_1$ (in the experimental tests reported in Sect. 3, $\bar{\rho}_1 = -0.1$ and $\bar{\rho}_2 = 0.1$). The versor $\alpha_v \in \mathbb{R}^3$ defines the postural synergy subspace of UB Hand (see our previous work [18]) and it has been obtained by computing $\alpha_v = (\alpha_{\text{closed}} - \alpha_{\text{open}})/(||\alpha_{\text{closed}} - \alpha_{\text{open}}||)$, being α_{open} and α_{closed} the hand postures in the synergy subspace that are closest to the hand configurations "totally open" and "totally closed". The matrix $S_P \in \mathbb{R}^{n_J \times n_S}$ is the robotic hand postural synergy matrix, where $n_S = 3$ is the number of considered *eigenpostures*, the reader may refer to [18] for the details on the computation of that matrix. Finally, the matrix

$$K_{\text{gain}} = \text{diag}(\underbrace{K_T, ..., K_T}_{\text{4 times}}, ..., \underbrace{K_i, ..., K_i}_{\text{4times}}, ..., \underbrace{K_P, ..., K_P}_{\text{4 times}}) \in \mathbb{R}^{n_J \times n_J}$$

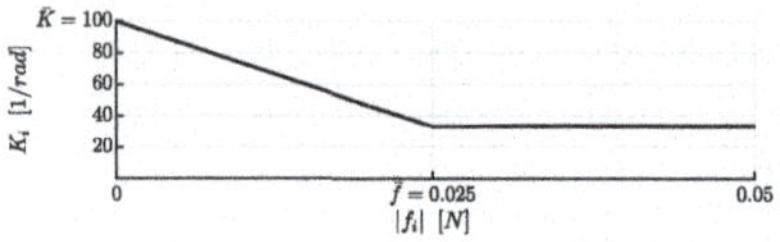

(a) Values of K_i as a function of the norm of the contact force.

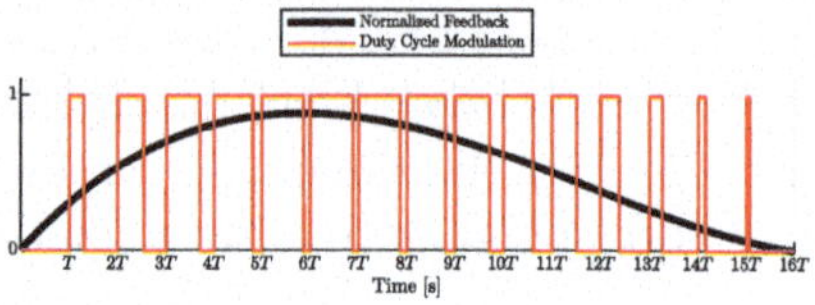

(b) Example of the duty-cycle modulation of the vibrotactile feedback stimulation.

Fig. 4. Values of K_i and duty-cycle modulation of the vibration motor.

Table 1. Objects used in the experiment.

Object Type	Box 	Cylinder 	Sphere 
Size [mm]	$88 \times 54 \times 54$	$100 \times 75 \times 75$	65 (*radius*)

where $i = \{T, I, M, R, P\}$, contains non-constant gains which are modulated by the contact force of each finger according to

$$K_i = \begin{cases} \bar{K}\left(\frac{1-\gamma}{\gamma}\frac{|f_i|}{\bar{f}} + 1\right), & \text{if } |f_i| < \bar{f} \text{ for } i = \{T, I, M\} \\ \dfrac{\bar{K}}{\gamma}, & \text{if } |f_i| \geq \bar{f} \text{ for } i = \{T, I, M\} \\ 0, & \forall |f_i| \quad\quad \text{for } i = \{R, P\} \end{cases}, \tag{11}$$

in which $|f_i| \in \mathbb{R}^+$ is the norm of the i-th finger force sensor, $\bar{f}$ is a constant threshold defining the minimum force value detected when the fingertip contacts an object, $\bar{K}$ is the control gain during free-space motions and γ is a scale factor determining the minimum value of K_i that may occur during the grasping. The variation of the gain K_i, that plays a central role in this control architecture, is motivated by a twofold reason: *(i)* for a given level of the myocontrol signal the motion of the finger that came into contact with the object is drastically slowed down, allowing the hand to adapt to the shape of the grasped object, and *(ii)* the capability of the user in finely modulating the grasp strength is highly increased, thanks to the possibility of providing small adjustment to the robotic hand configuration, see Eq. (9). Finally, note that, for $i = \{R, P\}$, K_i has been set to zero because in this study only thumb, index and middle fingers were equipped with tactile sensors and consequently only tripodal grasps have been considered. In Fig. 4(a) the value of K_i (for $i = \{T, I, M\}$) as a function of $|f_i|$ are shown.

2.4 Sensory Substitution for Contact Detection

Vibrotactile Feedback. In order to implement the robotic hand myocontrol, the user is provided with a vibrotactile stimulation produced by means of a vibration motor embedded on the wrist armband of the Noitom Hi5 glove [19] (see Fig. 3(c)), a wearable device for hand's finger motion tracking and feedback stimulation. Note that the motion tracking functionality of the glove is not used in this work, and only the vibration motor was applied to the wrist of the subjects during the experimental tests, as illustrated in Fig. 3. This vibration motor can be controlled via a Bluetooth interface at the maximum rate of 1 Hz by commanding the vibration duty cycle. Accordingly, the vibrotactile stimulus was obtained according to a *pulse-width modulation* strategy, in which the input signal $\nu \in [0, 1]$, hereafter called *normalized feedback signal*, determines the duty-cycle of the vibration motor activation signal, as illustrated in Fig. 4(b).

Force Sensing Interpretation. The vibrotactile stimulus is provided to the user on the basis of the normalized feedback signal ν, which depends on an estimation of internal forces generated by the grasp. Note that with internal force the forces exerted by the robotic hand fingers that produce zero resultant forces on the object is meant. Mathematically, these forces are defined as the force vector which spans the null space of the so-called *grasp matrix*, typically denoted with G [20]. In precision tripodal grasps, the thumb fingertip is located in an opposite configuration with respect to the index and middle fingertips, and its contact force is therefore counteracted by the other two fingers and viceversa. Consequently, the force exerted by the thumb at the contact location represents an good approximation of the applied internal force. For this reason, its measurement was used to approximate the internal force, and the normalized force signal was computed as

$$\nu = \begin{cases} \frac{\left||f_T| - \bar{f}_T\right|}{c_{\text{norm}}}, & \text{if } \left||f_T| - \bar{f}_T\right| \le c_{\text{norm}} \\ 0, & \text{if } \left||f_T| - \bar{f}_T\right| \le \beta \\ 1, & \text{if } \left||f_T| - \bar{f}_T\right| > c_{\text{norm}} \end{cases}, \tag{12}$$

where c_{norm} is a normalization constant and β denotes an error band in which the tracking of $\bar{f}_T$ is considered achieved (in our experiment $\beta = 0.03$, see Sect. 3.2). The combination of (12) with the duty cycle modulation of the vibration motor (Fig. 4(b)) produces the vibrotactile feedback. Note that, since the vibrotactile feedback depends on the absolute value of the error between estimated and target grasp strength, during the experiment reported in the following subsection the user was assisted by a simple visual cue (on a display) informing only about the sign of the error, see Fig. 5, while the extent of the error was available solely through the vibrotactile stimulation.

3 Experimental Tests

In order to assess the effectiveness of the proposed control architecture, an experiment is conducted involving four naïve able-bodied subjects. The test is performed in accordance with the Declaration of Helsinki.

3.1 Experimental Setup and Protocol

Figure 5 shows the setup arrangement used in the experimental session. The subjects are four men between 25 and 35 years old (namely S1, S2, S3, S4), and none of them experienced myocontrol before. During the experiment, each subject is seated in front of a desk equipped with the robotic Hand, described in Subsect. 2.3. A couple of differential sEMG electrodes are placed in proximity of the FDS and EDC muscles' belly (see Subsect. 2.2), and the vibration motor described in Subsect. 2.4 is fixed to the forearm in a location close to the wrist.

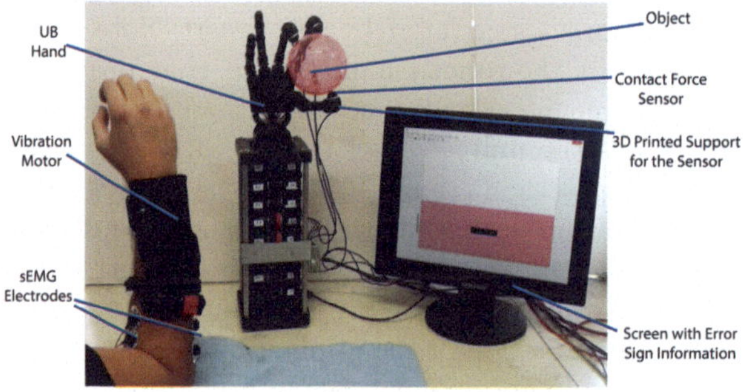

Fig. 5. The experimental setup for the grasp strength regulation experiment.

Initially, during the myocontrol training session, each subject is asked to fully open and close his hand two times while the sEMG signals are recorded. Then, the system automatically computed the pseudo-inverse of the muscular synergy matrix S_M^+ and the scaling constants as described in Sect. 2.2. After the myocontrol calibration, the grasping strength regulation task starts. When the myocontrol is connected to the UB Hand, the subject receive the vibrotactile feedback according to the strategy reported in Sect. 2. In each trial, the subject is initially requested to keep the robotic hand fully open, until the experimenter hands the first object out. At this point, the subject is asked to close the robotic hand in order to grasp the object with the three fingers, see Fig. 3. Once the object is grasped, the experimenter release the object and the subjects had to regulate – by exploiting the proposed myocontrol interface – the grasp strength, according to three given reference levels ($\bar{f}_{T,1} = 0.2\,\mathrm{N}$, $\bar{f}_{T,2} = 0.4\,\mathrm{N}$, $\bar{f}_{T,3} = 0.6\,\mathrm{N}$). Each of the grasping strength level has to be reached and maintained for a duration of approximately 100 s. Then the target level is changed *during* the grasp, i.e. without the subject releases the object. At the end, the experimenter vocally instructs the subject to release the object (i.e. open the robotic hand) and take the object back. Three different objects are used during the experiments: a box, a cylinder and a sphere (supplied in this order), whose dimensions and images are reported in Table 1. Three different grasping and regulation tasks are carried out for the three different objects. The subject rests 10 min after each task.

The results of the experiment are reported in the following paragraph, using as metrics the mean and standard deviation of the grasp strength regulation profiles over the subjects, and the related settling times as an index of success in the task execution.

3.2 Grasp Strength Regulation Results

Figure 6 shows the grasping strength signal obtained during the regulation task of one of the subjects (S4) with the Box object. The graph at the top shows that after a transient the prescribed grasp strength – well represented by the thumb force $|f_T|$ – was successfully applied by the user regardless of the specific level. Note the tracking of a reference level is considered successfully achieved when $|f_T|$ enters within an error band equal to $\pm 5\%$ of the higher strength reference level (0.6 N), i.e. in the range $[\bar{f}_{T,i} - 0.03, \bar{f}_{T,i} + 0.03]$, $i = 1, 2, 3$, and remains in this range until a new reference level is provided or until the experiment ends. The settling time is defined as the time required by the user to successfully reach the desired grasp strength level.

By focusing on the graph showing the hand closure level σ_{input} in Fig. 6, it is possible to observe how this signal was intuitively regulated in order to increase (time instants around 0 s, 100 s, 200 s) or decrease the force (time instants around 30 s, 290 s). This is more clearly shown in the graph plotting σ, the myocontrol-driven signal given by Eq. (10), where the values inside the deadzone are set to zero.

Finally, the bottom plot of Fig. 6 illustrates how σ is modulated by K_{gain}: when $|f_T| < \bar{f}$ (orange dashed line in the top graph of Fig. 6), K_T assumes the highest value, see the *case#1* in Eq. (11), while when $f_T \geq \bar{f}$, i.e. contact with the object is established, the gain K_T quickly decreases to its minimum level equal to $\bar{K}/\gamma$ according to *case#2* of Eq. (11). Accordingly, the signal $K_T\,\sigma$ becomes smaller and smaller during the interaction with the object, and therefore the user capability of modulating the robotic hand joint velocity references becomes more and more accurate.

The experiments highlight that the behaviour shown in Fig. 6 is replicated by all the subjects, since all of them successfully regulate the grasping force for all the considered reference levels. These results are shown in Fig. 7 in terms of mean and standard deviation values of the regulated grasp strength, computed over the subjects. Note that, even if the performances of the users during the transient are quite different, producing a remarkable variability of the force strength during this phase, the desired final value is achieved by all the subjects – in terms of successful reference tracking, as defined according to the settled time previously

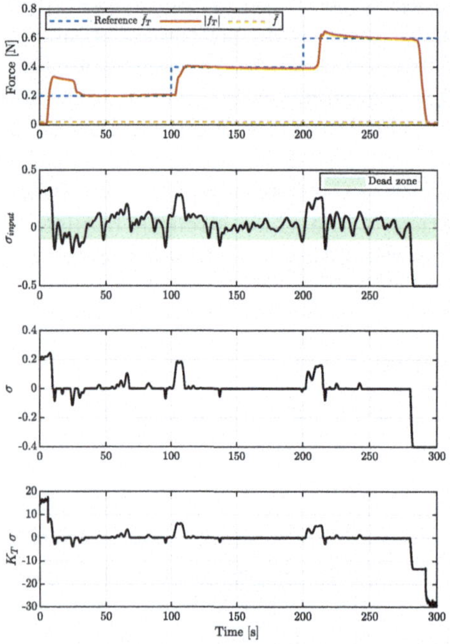

Fig. 6. Signals of the strength regulation with the Box object (sbj. S4).

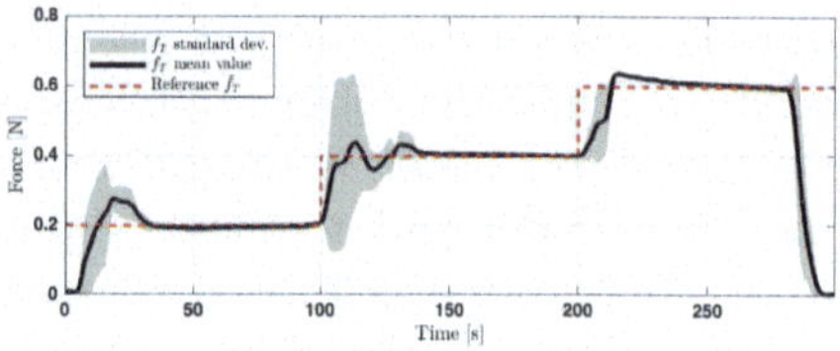

(a) Box object.

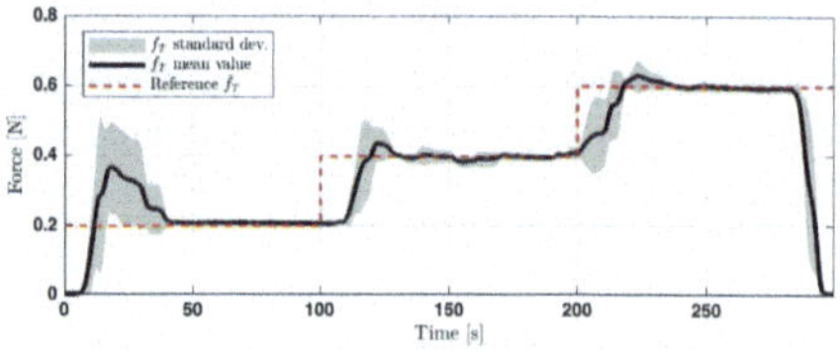

(b) Cylinder object.

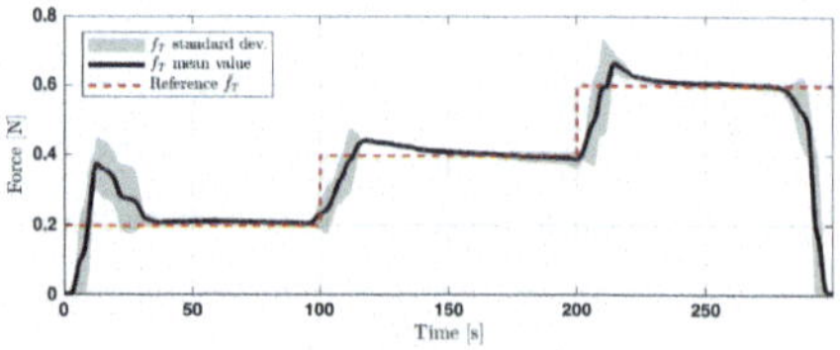

(c) Sphere object.

Fig. 7. Mean and standard deviation of the regulated grasp strength, computed over the subjects.

mentioned in this paragraph –, as shown by the mean and standard deviation of the regulated grasp strengths reported in Fig. 7.

4 Conclusions

In this paper, the implementation and the experimental evaluation of a human machine interface based on a synergy mapping of sEMG measurements and vibrotactile feedback for the regulation of the grasping strength on an anthropomorphic robotic hand is reported. The experiments demonstrate that the proposed framework allow a fast adaptation to different subjects thanks to a simple and fast online calibration procedure. Moreover, the synergy-based sEMG mapping simplified the detection of the human intention, enabling the implementation of a reliable interface between the human and the robotic hand, at least for a simple tripodal grasping task. The experimental results highlighted also that, thanks to the vibrotactile feedback, all the subjects are able to properly regulate different values of desired grasping strength.

Despite the simplicity of the adopted feedback strategy, the results reported in this paper are very promising and pave the way toward the extension of the proposed approach to a wider set of tasks and more advanced and flexible feedback techniques. Therefore, future work will be devoted to the experimental validation of the approach by extending the set of considered grasps and in the improvement and evaluation of the proposed vibrotactile feedback, as well as on the investigation of alternative feedback strategies. Finally, a larger set of subjects will be engaged for future experiments, in order to increase the experimental protocol's statistical power and provide an analysis of the statistically significance level of the results for a proper discussion on the generalization of the reported outcomes.

References

1. Santello, M., Bianchi, M., Gabiccini, M., Ricciardi, E., Salvietti, G., Prattichizzo, D., Ernst, M., Moscatelli, A., Jörntell, H., Kappers, A.M., et al.: Hand synergies: integration of robotics and neuroscience for understanding the control of biological and artificial hands. Phys. Life Rev. **17**, 1–23 (2016)
2. Santello, M., Flanders, M., Soechting, J.F.: Postural hand synergies for tool use. J. Neurosci. **18**(23), 10105–10115 (1998)
3. Matsumura, Y., Mitsukura, Y., Fukumi, M., Akamatsu, N., Yamamoto, Y., Nakaura, K.: Recognition of EMG signal patterns by neural networks. In: Proceedings of the 9th International Conference on Neural Information Processing, ICONIP 2002, vol. 2, pp. 750–754. IEEE (2002)
4. Young, A.J., Smith, L.H., Rouse, E.J., Hargrove, L.J.: Classification of simultaneous movements using surface emg pattern recognition. IEEE Trans. Biomed. Eng. **60**(5), 1250–1258 (2013)
5. Liarokapis, M.V., Artemiadis, P.K., Katsiaris, P.T., Kyriakopoulos, K.J., Manolakos, E.S.: Learning human reach-to-grasp strategies: Towards EMG-based control of robotic arm-hand systems. In: 2012 IEEE International Conference on Robotics and Automation (ICRA), pp. 2287–2292. IEEE (2012)
6. Jiang, N., Englehart, K.B., Parker, P.A.: Extracting simultaneous and proportional neural control information for multiple-dof prostheses from the surface electromyographic signal. IEEE Trans. Biomed. Eng. **56**(4), 1070–1080 (2009)
7. Bach-y Rita, P., Kercel, S.W.: Sensory substitution and the human-machine interface. Trends Cogn. Sci. **7**(12), 541–546 (2003)
8. Cipriani, C., Zaccone, F., Micera, S., Carrozza, M.C.: On the shared control of an EMG-controlled prosthetic hand: analysis of user-prosthesis interaction. IEEE Trans. Robot. **24**(1), 170–184 (2008)
9. Pasluosta, C., Tims, H., Chiu, L., et al.: Slippage sensory feedback and nonlinear force control system for a low-cost prosthetic hand. Am. J. Biomed. Sci. **1**(4), 295–302 (2009)
10. Saunders, I., Vijayakumar, S.: The role of feed-forward and feedback processes for closed-loop prosthesis control. J. Neuroeng. Rehabil. **8**(1), 60 (2011)
11. Rohland, T.A.: Sensory feedback in upper limb prosthetic systems. Inter Clin. Inf. Bull. **13**(1), 1–4 (2013)
12. Benatti, S., Casamassima, F., Milosevic, B., Farella, E., Schönle, P., Fateh, S., Burger, T., Huang, Q., Benini, L.: A versatile embedded platform for emg acquisition and gesture recognition. IEEE Trans. Biomed. Circuits Syst. **9**(5), 620–630 (2015)

13. Burdet, E., Ganesh, G., Yang, C., Albu-Schäffer, A.: Interaction force, impedance and trajectory adaptation: by humans, for robots. In: Experimental Robotics, pp. 331–345. Springer (2014)
14. Melchiorri, C., Palli, G., Berselli, G., Vassura, G.: Development of the UB Hand IV: overview of design solutions and enabling technologies. IEEE Robot. Autom. Mag. **20**(3), 72–81 (2013)
15. Palli, G., Melchiorri, C., Vassura, G., Scarcia, U., Moriello, L., Berselli, G., Cavallo, A., De Maria, G., Natale, C., Pirozzi, S., et al.: The DEXMART hand: mechatronic design and experimental evaluation of synergy-based control for human-like grasping. Int. J. Robot. Res. **33**(5), 799–824 (2014)
16. OnRobot OptoForce OMD-20-SE-40N (2019). https://onrobot.com/en
17. Robotis Corporation (2016). http://www.en.robotis.com/
18. Ficuciello, F., Palli, G., Melchiorri, C., Siciliano, B.: Experimental evaluation of postural synergies during reach to grasp with the UB Hand IV. In: 2011 IEEE/RSJ International Conference on Intelligent Robots and Systems, pp. 1775–1780. IEEE (2011)
19. Noitom Hi5 VR Glove (2019). https://www.noitom.com/solutions/hi5-vr-glove
20. Murray, R.M.: A Mathematical Introduction to Robotic Manipulation. CRC Press, Boca Raton (2017)

Hierarchical Task-Priority Control for Human-Robot Co-manipulation

Jonathan Cacace, Fabio Ruggiero(✉), and Vincenzo Lippiello

PRISMA Lab, Department of Electrical Engineering and Information Technology, University of Naples Federico II, Via Claudio 21, 80125 Naples, Italy
{jonathan.cacace,fabio.ruggiero,vincenzo.lippiello}@unina.it

Abstract. The extensive distribution of collaborative robots in industrial workplaces allows human operators to decrease the weight and the repetitiveness of their activities. In order to facilitate the role of the human worker during the interaction with these robots, innovative control paradigms, enabling an intuitive human-robot collaborative manipulation, are needed. In this work, a dynamic and hierarchical task-priority control framework is proposed, leveraging a physical interaction with a redundant robot manipulator through a force sensor. The foremost objective of this approach is to exploit the non-trivial null-space of the redundant robot to increase the performance of the co-manipulation and, consequently, its effectiveness. A comparison between the proposed methodology and a standard admittance control scheme is carried out within an industrial use case study consisting of a human operator interacting with a KUKA LBR iiwa arm.

Keywords: Task-priority control · Human-robot interaction · Cobots

1 Introduction

Collaborative human-robot manipulation is an emerging field pointing at developing innovative service robotic applications in which humans and robots cooperate for the execution of tasks sharing their workspace, as highlighted by Corrales et al. (2012). Thanks to the extensive distribution of collaborative robot, also known as *cobots*, a robotic co-worker can assist human operators during the execution of service tasks, like cooperative lifting and transportation of heavy objects, collaborative assembly, and similar. The assistance can be carried out both in an active way, as underlined by Caccavale et al. (2016), and in a proactive way, as explained by Zanchettin and Rocco (2017). In this scenario, the human worker typically guides the motion of the robot by physically interacting with its end-effector, whose reaction is compliant to the forces exerted by the human.

In order to operate in safe and adequate conditions, the robot motion should be controlled by taking into account the performance of the manipulator and avoiding instability regions, kinematic singularities, and joint mechanical limitations. If one of these situations occurs, the robot loses the ability to move since the operator forces cannot be translated into suitable joint commands.

F. Ferraguti et al. (Eds.): HFR 2019, SPAR 12, pp. 125–138, 2020.
https://doi.org/10.1007/978-3-030-42026-0_10

In this work, the kinematic redundancy of the manipulator is exploited to design a set of prioritized tasks. These tasks involve the forces exerted by the operator to move the robot accordingly. Such forces are measured through a force sensor mounted at the end-effector. Besides, the tasks also involve several indices related to the manipulator performance such as the robot manipulability, the joint limits, and other quality factors as described by Chiaverini (1997) and Baerlocher and Boulic (1998). While the human operator moves the manipulator end-effector through the force sensor, the proposed formulation maximizes the performance of the robot during such cooperative manipulation. Since the input issued by the human operator is not predictable, the robot motion cannot be optimally planned in advance. In this context, several factors, such as the operator inexperience, fatigue, or stress, may bring the manipulator in a low-performance configuration. As a result, the robot can get stuck or, in worst cases, it becomes unstable and dangerous for a tight interaction with the human. Therefore, robot motion must be adapted at runtime. The task priority formulation can exploit the degrees of freedom (DoFs) needed to accomplish the primary task (e.g., follow the movements of the human operator given through the force sensor) while relying on the non-trivial null-space of the redundant robot, and thus on the remaining DoFs, to satisfy secondary tasks. Besides, in the proposed formulation, dynamic switching of the task priority is proposed to modify the stack structure (e.g., change primary goal and, consequently, the null-space composition of tasks).

The proposed framework is validated through a human-robot interactive task where a human operator cooperates with a KUKA LBR iiwa robot on whose end-effector is mounted a force sensor, as shown in Fig. 1. The effectiveness of the system is also evaluated by comparing the proposed methodology with a standard admittance controller.

The outline of the paper is organized as follows. A brief overview of the related works is presented in Sect. 2, in which the proposed contribution is pointed out. The hierarchical task priority framework used to implement the null-space control for co-manipulation is described in Sect. 3. The employed tasks are introduced in Sect. 4, while in Sect. 5 their composition is discussed. Finally, in Sect. 6, an experimental case study to test the effectiveness of the approach is carried out. Section 7 concludes the paper through a brief discussion of the overall work and future research directions.

2 Related Works

Within human-robot co-manipulation systems, human operators can usually physically interact with a robot by applying forces on its end-effector directly. In these applications, the operator forces are translated into robot motion commands imposing a desired dynamic behavior between the external forces and the robot motion: this is possible thanks to an admittance control scheme, for instance. Several works addressed the development of intuitive interaction strategies to improve the performance of the co-manipulation starting from the fundamental works of Newman (1992) and Gullapalli et al. (1992).

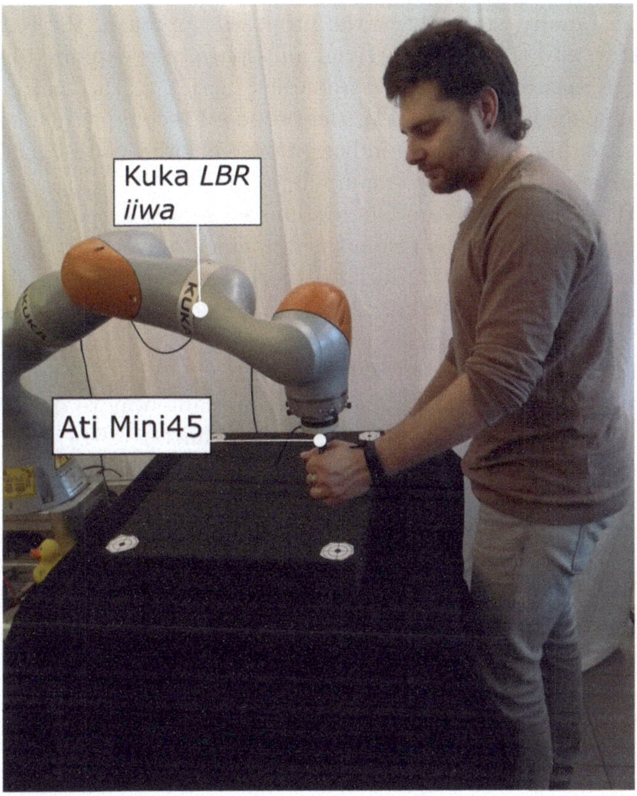

Fig. 1. The human operator physically interacts with a lightweight manipulator through the *ATI Mini45* force sensor.

Different indices are proposed in the literature to evaluate the performance of the manipulator in a given configuration. Yoshikawa (1985) introduced a manipulability index, while Dubey and Luh (1988) introduced the end-effector velocity ratio (*MVR*) index. Landi et al. (2017) presented a parameter to calculate the deviations of the manipulator from the nominal behavior, and therefore identify control vulnerabilities.

Several works addressed the problem of increasing manipulator performance during human-robot co-manipulation tasks. The following approaches can be distinguished. Within variable admittance approaches, the gains of the admittance controller are regulated on-line to provide an intuitive interaction between the robotic arm and the operator. In particular, Labrecque and Gosselin (2018) regulated the controller gains to accommodate and optimize the system response to any potential dynamics, while Landi et al. (2017) adjusted the damping of the admittance controller to avoid undesired and unstable robot configurations. Dimeas et al. (2018) proposed instead a haptic feedback approach to prevent the operator from bringing the manipulator in a low-performance region of the workspace. Finally, similarly to the proposed approach, the redundancy of the manipulator, and thus its non-trivial null-space, was exploited by Ficuciello et al. (2014) to ensure a decoupled apparent inertia at the end effector. Differently from Ficuciello et al. (2014), the proposed technique defines a stack of tasks to optimize different quality indices, whose priority can be modified based on the current behavior of the co-manipulation task.

Task priority control of redundant systems is anyway extensively explored in the literature. Applications span from the fundamental work by Nakamura et al. (1987) to more recent utilization in fine object manipulation, aerial and underwater scenarios by Caccavale et al. (2013), Lippiello et al. (2016), and Casalino et al. (2012), respectively. Task composition was exploited by Rocchi et al. (2015) and Mansard et al. (2009) to control collaborative working of humanoid robots, while a framework providing a flexible and composable way to define robot control tasks was proposed by Aertbeliën and De Schutter (2014). Finally, in the context of collaborative applications and similarly to the proposed approach, Agravante et al. (2014) designed a task composition methodology coupling a visual servoing and an impedance controller. Differently from what proposed in this work, the compliant behaviour does not seem introduced to accommodate the human control input, but to achieve the desired collaborative task. Therefore, at the best of authors' knowledge, no application in which the human operator physically leads the motion of the robot can be found with hierarchical task-priority control. Hence, the novelty introduced by this work is indeed the application of the renowned null-space task priority formulation within a cobot system.

3 Dynamic Task Priority Framework

In this section, the dynamic task priority framework used to implement the co-manipulation system is explained. A similar formulation was already introduced by Buonocore et al. (2015) and Lippiello et al. (2016) in different application domains.

Let n be the number of DoFs of the robot manipulator. To exploit kinematic redundancy of the system, the dimension μ_0 of the main task for the robot end-effector must be less than n. In this way, it is possible to achieve some secondary tasks in terms of priority, while preserving the principal task, exploiting the extra DoFs. Hence, in a redundant manipulator, the same position of the end-effector can be reached through different joint configurations, letting the robotic arm to be suitably reconfigured by using internal motions.

Let $\boldsymbol{q} = \begin{bmatrix} q_1 \dots q_n \end{bmatrix}^\top \in \mathbb{R}^n$ be the arm joint vector describing the arm configuration, where q_i is the i-th joint variable, with $i = 1, \dots, n$. The goal of the kinematic control framework is to accomplish the primary and, eventually, the secondary tasks generating the control velocity vector $\dot{\boldsymbol{q}} \in \mathbb{R}^n$ (i.e., the velocity command of the robot joints).

Let $\boldsymbol{\sigma}_0 := \boldsymbol{f}_0(\boldsymbol{q}) \in \mathbb{R}^{\mu_0}$ be the main task. The following differential relationship holds:

$$\dot{\boldsymbol{\sigma}}_0 := \frac{\partial \boldsymbol{f}_0(\boldsymbol{q})}{\partial \boldsymbol{q}}\dot{\boldsymbol{q}} = \boldsymbol{J}_0(\boldsymbol{q})\dot{\boldsymbol{q}}, \tag{1}$$

where $\boldsymbol{J}_0(\boldsymbol{q}) \in \mathbb{R}^{\mu_0 \times n}$ represents the main task Jacobian. In order to fulfill the regulation problem of $\boldsymbol{\sigma}_0$ to the desired value $\boldsymbol{\sigma}_0^*$, let $\widetilde{\boldsymbol{\sigma}}_0 = \boldsymbol{\sigma}_0^* - \boldsymbol{\sigma}_0$ be the main task error. By inverting (1) the following velocity command can be considered

$$\dot{\boldsymbol{q}}^* = \boldsymbol{J}_0(\boldsymbol{q})^\dagger \boldsymbol{\Lambda}_0 \widetilde{\boldsymbol{\sigma}}_0, \tag{2}$$

where $\boldsymbol{\Lambda}_0 \in \mathbb{R}^{\mu_0 \times \mu_0}$ is a positive-definite gain matrix, and $\boldsymbol{J}_0(\boldsymbol{q})^\dagger$ is the generalized pseudo-inverse of $\boldsymbol{J}_0(\boldsymbol{q})$. The following error dynamics is achieved folding (2) into (1)

$$\dot{\widetilde{\boldsymbol{\sigma}}}_0 = -\boldsymbol{\Lambda}_0 \widetilde{\boldsymbol{\sigma}}_0, \tag{3}$$

obtaining an asymptotic convergence of the main task to its desired value.

Since $\mu_0 < n$, the null-space of $\boldsymbol{J}_0(\boldsymbol{q})$ is non-trivial. A further task can be thus defined as $\boldsymbol{\sigma}_1 := \boldsymbol{f}_1(\boldsymbol{q}) \in \mathbb{R}^{\mu_1}$, with $\mu_1 \leq n$, from which the following joint velocity command can be designed

$$\dot{\boldsymbol{q}}^* = \boldsymbol{J}_0(\boldsymbol{q})^\dagger \boldsymbol{\Lambda}_0 \widetilde{\boldsymbol{\sigma}}_0 + \boldsymbol{N}_0(\boldsymbol{q}) \boldsymbol{J}_1(\boldsymbol{q})^\dagger \boldsymbol{\Lambda}_1 \tilde{\boldsymbol{\sigma}}_1, \tag{4}$$

where $\boldsymbol{J}_1(\boldsymbol{q}) \in \mathbb{R}^{\mu_1 \times n}$ is the Jacobian matrix of the defined task and it is built similarly to (1), $\tilde{\boldsymbol{\sigma}}_1$ is the error with respect to a desired value $\boldsymbol{\sigma}_1^*$ for the task, $\boldsymbol{\Lambda}_1 \in \mathbb{R}^{\mu_1 \times \mu_1}$ is a positive-definite gain matrix, and $\boldsymbol{N}_0(\boldsymbol{q}) = \boldsymbol{I}_n - \boldsymbol{J}_0(\boldsymbol{q})^\dagger \boldsymbol{J}(\boldsymbol{q})$ is the projector into the null-space of $\boldsymbol{J}_0(\boldsymbol{q})$, with $\boldsymbol{I}_n \in \mathbb{R}^{n \times n}$ the identity matrix.

Adding further secondary tasks, Antonelli (2009) generalizes the framework to the case of η prioritized tasks with the following general velocity command

$$\dot{\boldsymbol{q}}^* = \boldsymbol{J}_0(\boldsymbol{q})^\dagger \boldsymbol{\Lambda}_0 \widetilde{\boldsymbol{\sigma}}_0 + \sum_{i=1}^{\eta} \boldsymbol{N}_{0|\ldots|i-1}(\boldsymbol{q}) \boldsymbol{J}_i(\boldsymbol{q})^\dagger \boldsymbol{\Lambda}_i \widetilde{\boldsymbol{\sigma}}_i, \tag{5}$$

where $\boldsymbol{N}_{0|\ldots|i}(\boldsymbol{q})$ is the projector onto the null space of the *augmented Jacobian* $\boldsymbol{J}_{0|\ldots|i}(\boldsymbol{q})$ of the i-th task, with $i = 0, \ldots, \eta - 1$, which are respectively defined as follows

$$\boldsymbol{J}_{0|\ldots|i}(\boldsymbol{q}) = \left[\boldsymbol{J}_0(\boldsymbol{q})^\top \cdots \boldsymbol{J}_i(\boldsymbol{q})^\top\right]^\top \tag{6}$$

$$\boldsymbol{N}_{0|\ldots|i}(\boldsymbol{q}) = \boldsymbol{I}_n - \boldsymbol{J}_{0|\ldots|i}(\boldsymbol{q})^\dagger \boldsymbol{J}_{0|\ldots|i}(\boldsymbol{q}). \tag{7}$$

The proposed priority-based task formulation (5) ensures that the execution of the main task is not affected by the remaining tasks in the assigned order. In other words, the execution of the lower priority tasks is subordinated to the execution of the higher priority ones. In this context, the lower priority task will be satisfied if enough and qualified DoFs are available, while the complete fulfillment of the main task is instead always guaranteed. Notice that the convergence of the task errors depends on the annihilation and independence properties of the task Jacobians as defined by Antonelli (2009).

3.1 Dynamic Modification of the Task Composition

As anticipated within the sections above, the considered tasks are put in a prioritized stack that can be on-line modified either by insert/removing a task in/from the stack, or by switching the priority order of one or more tasks. In order to address possible discontinuities of the control input that could be generated by the modification of the stack, a time-vanishing smoothing term is adopted to manage the transitions. Specifically, suppose that the transition phase starts at $t = 0$ and the r-th task of the stack must be deactivated and substituted by the $(r+1)$-th task. During the transition, the velocity command is computed like Caccavale et al. (2013) as

$$\dot{\boldsymbol{q}}^*(t) = \dot{\boldsymbol{q}}^*_{r+1}(t) + e^{-\frac{t}{\tau}} \left(\dot{\boldsymbol{q}}^*_r(0) - \dot{\boldsymbol{q}}^*_{r+1}(0) \right), \tag{8}$$

where τ is a positive time constant determining the transition phase duration, and $\dot{\boldsymbol{q}}^*_k$ is the velocity command corresponding to the k-th task of the stack. When t becomes sufficiently higher than τ, the r-th task stack is entirely removed, and a new transition can start. Similarly, Eq. (8) can be employed to switch the priority between two tasks in the stack. Consider that Eq. (8) only affects the computed commanded velocity during the transition of different stacks after their composition. The main aim of this equation is to make smoother the activation/deactivation of given tasks.

4 Definition of Possible Tasks for a Cobot System

In this section, the possible tasks of interest for a cobot system are presented. In particular, two different kinds of tasks are discussed: positioning and performance tasks. The latter group includes all the tasks affecting the arm manipulability, while the former tasks are related to the control of the position and orientation of the robot end-effector.

4.1 End-Effector Position Task

The goal of this task is to hold the position of the robot end-effector in a desired configuration. Let $\boldsymbol{J}(\boldsymbol{q}) \in \mathbb{R}^{6\times n}$ be the Jacobian matrix of the robotic arm, which can be defined as follows

$$\boldsymbol{J}(\boldsymbol{q}) = \begin{bmatrix} \boldsymbol{J}_p(\boldsymbol{q}) \\ \boldsymbol{J}_o(\boldsymbol{q}) \end{bmatrix}, \tag{9}$$

where $\boldsymbol{J}_p(\boldsymbol{q}) \in \mathbb{R}^{3\times n}$ and $\boldsymbol{J}_o(\boldsymbol{q}) \in \mathbb{R}^{3\times n}$ represent the parts of the Jacobian mapping the joint velocities to the linear and rotational velocities of the end-effector, respectively. Let $\boldsymbol{p}(\boldsymbol{q}) \in \mathbb{R}^3$ and $\boldsymbol{p}^* \in \mathbb{R}^3$ be the current and desired position of the end-effector in the robot base frame, respectively. The corresponding task function is defined as $\boldsymbol{\sigma}_p = \boldsymbol{p}^* - \boldsymbol{p}(\boldsymbol{q})$. Hence, in order to fulfill the task, the reference $\boldsymbol{\sigma}^*_p = \boldsymbol{0}$ is chosen. The corresponding Jacobian matrix for the end-effector positioning task is $\boldsymbol{J}_p(\boldsymbol{q})$.

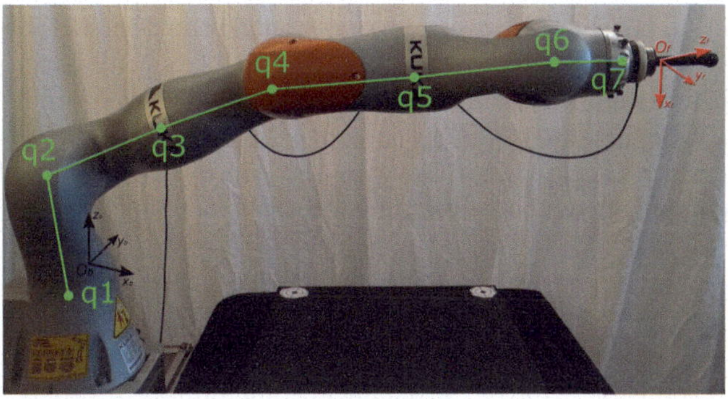

Fig. 2. KUKA LBR iiwa robot joints depicted in green. In black, the robot base frame. In red, the end-effector frame.

4.2 End-Effector Orientation Task

The purpose of this task is to control the orientation of the robot end-effector. In the proposed scenario, the robot should maintain its orientation in a configuration convenient with the considered co-manipulation task. For instance, considering the proposed setup, the end-effector of the robot is commanded to maintain the robot flange facing downwards, as depicted in Fig. 1. Let $\boldsymbol{R}^* \in SO(3)$ and $\boldsymbol{R}(\boldsymbol{q}) \in SO(3)$ be the rotation matrices representing the desired and current orientation of the robot end-effector, respectively. The corresponding unit quaternions are denoted by $\{\eta_o(\boldsymbol{q}), \boldsymbol{\epsilon}_o(\boldsymbol{q})\}$ and $\{\eta_o^*, \boldsymbol{\epsilon}_o^*\}$, respectively. The orientation error can be computed as follows

$$\boldsymbol{e}_o(\boldsymbol{q}) = \eta_o(\boldsymbol{q})\boldsymbol{\epsilon}_o^* - \eta_o^*\boldsymbol{\epsilon}_o(\boldsymbol{q}) - \boldsymbol{S}(\boldsymbol{\epsilon}_o^*)\boldsymbol{\epsilon}_o(\boldsymbol{q}), \tag{10}$$

where $\boldsymbol{S}(\cdot)$ is the skew-symmetric matrix. With reference to Fig. 2, the sought goal is to keep the z-axis of the end-effector facing downwards, neglecting the rotation around such an axis. Therefore, only the first two components of the error vector calculated in (10) are employed in the following experiments. Hence, the task function is designed as $\boldsymbol{\sigma}_o = \boldsymbol{\Sigma}\boldsymbol{e}_o$, with $\boldsymbol{\Sigma} = \begin{bmatrix} 1 & 0 & 0 \\ 0 & 1 & 0 \end{bmatrix}$, while the desired task variable is $\boldsymbol{\sigma}_o^* = \boldsymbol{0}$. The corresponding task Jacobian matrix is $\boldsymbol{J}_o' = \boldsymbol{\Sigma}\boldsymbol{J}_o \in \mathbb{R}^{2\times n}$.

4.3 Interaction Force Task

This task intends to implement the compliance behavior of the manipulator exploiting the data generated by the force sensor. Roughly speaking, this task pursues a zero-force control allowing the position of the robot end-effector to be freely moved by the human operator.

Let $\boldsymbol{f}^s \in \mathbb{R}^3$ be the interaction force vector in the sensor frame and $\boldsymbol{R}_s(\boldsymbol{q}) \in \mathbb{R}^{3\times 3}$ the rotation matrix between the sensor frame and the robot base frame.

The task function is defined as $\boldsymbol{\sigma}_f = \boldsymbol{R}_s(\boldsymbol{q})\boldsymbol{f}^s$, while the desired value is $\boldsymbol{\sigma}_f^* = \boldsymbol{0}$. In this way, the manipulator moves the position of its end-effector until the human operator does not interact with the robot. The corresponding Jacobian matrix of the task is $\boldsymbol{J}_p(\boldsymbol{q}) - \boldsymbol{S}(\boldsymbol{p}_s(\boldsymbol{q}) - \boldsymbol{p}(\boldsymbol{q}))\boldsymbol{J}_o(\boldsymbol{q})$, with $\boldsymbol{p}_s(\boldsymbol{q}) \in \mathbb{R}^3$ is the position of the tip of the sensor force in the robot base frame. However, to simplify the computations, the approximated task Jacobian $\boldsymbol{J}_p(\boldsymbol{q})$ is considered for this task.

4.4 Joint Limits Avoidance Task

This task is responsible to avoid mechanical joint limits. In order to bring the joints as far as possible from their limits, the robot is solicited to keep a configuration in the middle of such a range. Therefore, the desired joints configuration can be computed as $\boldsymbol{q}^* = \boldsymbol{q}_L + \frac{1}{2}(\boldsymbol{q}_H - \boldsymbol{q}_L)$, where $\boldsymbol{q}_L \in \mathbb{R}^n$ and $\boldsymbol{q}_H \in \mathbb{R}^n$ are the minimum and maximum mechanical limits, respectively. The task function σ_l corresponding to the joint limits task is defined as

$$\sigma_l(\boldsymbol{q}) = \frac{1}{2n}\sum_{i=1}^{n}\left(\frac{q_i - q_i^*}{q_{H_i} - q_{L_i}}\right)^2, \tag{11}$$

with q_i^*, q_{H_i}, and q_{L_i} being the i-th component of $\boldsymbol{q}^*$, $\boldsymbol{q}_H$ and $\boldsymbol{q}_L$, respectively. Equation (11) represents a weighted sum of the error of each joint with respect to its desired configuration in the middle of the mechanical range. The desired value for the task is $\boldsymbol{\sigma}_l^* = 0$, while the corresponding Jacobian matrix $\boldsymbol{J}_l \in \mathbb{R}^{1\times n}$ is equal to

$$\boldsymbol{J}_l = \frac{1}{n}\left[\frac{q_1 - q_1^*}{(q_{H_1} - q_{L_1})^2} \cdots \frac{q_n - q_n^*}{(q_{H_n} - q_{L_n})^2}\right]. \tag{12}$$

4.5 Manipulability Task

The goal of this task is to prevent the arm to reach kinematic singularity configurations. For this scope, the manipulability index introduced in Siciliano et al. (2009) might be considered. Given an arbitrary arm configuration $\boldsymbol{q}$, the manipulability index is calculated as

$$w(\boldsymbol{q}) = \sqrt{\det\left(\boldsymbol{J}(\boldsymbol{q})\boldsymbol{J}(\boldsymbol{q})^\top\right)}. \tag{13}$$

However, to save computational resources, simpler manipulability indices exist. In this work, the solution adopted by Caccavale et al. (2013) is considered. The manipulability task function can be thus designed in the following way

$$\sigma_m(\boldsymbol{q}) = \frac{1}{2}\sum_{i=2}^{n-1}\sin^2(q_i). \tag{14}$$

The first and last joints are neglected for the considered robot since they do not cause singularity issues. Notice that, for the considered robot manipulator, the

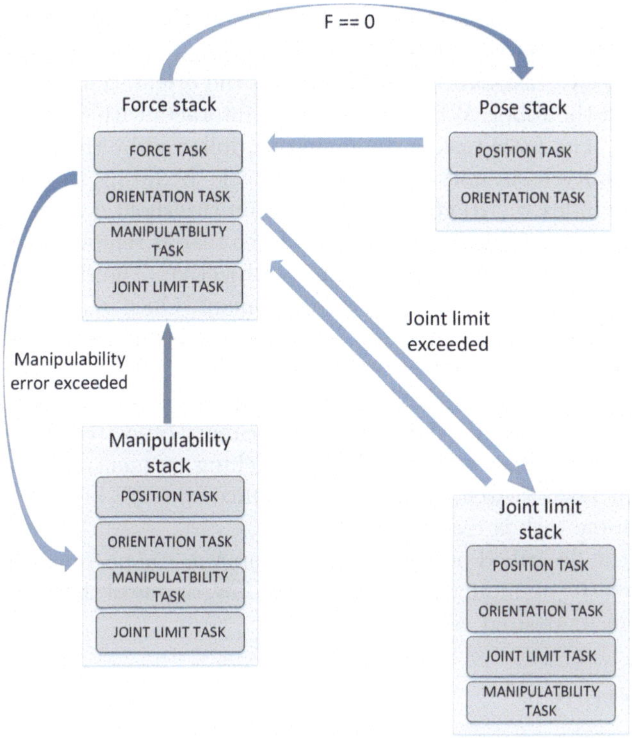

Fig. 3. Policy employed for the addressed co-manipulation system. Four stacks are represented, each of one contains a prioritized order of tasks. The corresponding condition regulates the switch between the stacks. When the switching condition does not hold anymore, the force stack is restored.

above equation has the same local minima of (13), with low values when the robot configuration is close to kinematic singularities.

The desired task value is $\boldsymbol{\sigma}_m^*$, which is experimentally tuned on the basis of the available setup. One way to retrieve $\boldsymbol{\sigma}_m^*$ is to compute (14) on a discretization of the workspace and take the mean value on this set. Otherwise, it is possible to retrieve $\boldsymbol{\sigma}_m^*$ from a preferred dexterous manipulator configuration. Finally, the corresponding Jacobian matrix $\boldsymbol{J}_m(\boldsymbol{q}) \in \mathbb{R}^{1 \times n}$ for the task is calculated as

$$\boldsymbol{J}_m(\boldsymbol{q}) = \left[\cos(q_1)\sin(q_1) \dots \cos(q_n)\sin(q_n)\right]. \quad (15)$$

5 Dynamic Task Composition

As previously declared, the stack of the active tasks during the co-manipulation action can be modified on-line by activating or deactivating a certain number of tasks or by changing their priority. The policy employed in the carried out experiments is represented in Fig. 3. Notice that, in principle, any combination of tasks and priorities can be adopted based on the desired cobot behavior. When the human operator does not exert any force on the robot end-effector, the only two active tasks are the position and the orientation ones, keeping the reached pose of the robot. When the operator interacts with the robot, the force stack is active. The force task is the main one, followed by the orientation task, the manipulability task, and the joint limit task. If during the interaction the manipulability index $\sigma_m(\boldsymbol{q})$ goes under a certain threshold $\bar{\sigma}_m$, the manipulability stack is activated. This means that the force task is removed from the stack, and the position task becomes the primary objective. The desired end-effector position is set as the one at the switching moment. Once $\sigma_m(\boldsymbol{q})$ becomes again greater than $\bar{\sigma}_m$, the force task is restored. Instead, if during the interaction the following inequality $|q_i - q_i^*| > \bar{\sigma}_l$ holds for a joint i, with $\bar{\sigma}_l$ a given threshold, the joint limit stack is activated. This means that the force task is removed from the stack, and the position task becomes the primary objective. The desired end-effector position becomes the one at the switching moment. The manipulation and joint limit tasks switch their priorities. Once $|q_i - q_i^*|$ becomes again lower than $\bar{\sigma}_l$, the force task is restored.

Through the described policy, the human operator is prevented from moving the robot when the manipulability index is too low, or one of the joint is diverging too much from its desired value.

6 Case Study

The effectiveness of the proposed system has been assessed by defining a collaborative manipulation case study where a human worker cooperates with a lightweight redundant robotic arm to bring its end-effector in predefined locations. In order to show the advantage of the presented approach, a user, already trained on the system, performed a test with the proposed method and one with a classical admittance control scheme. A video of one representative experiment can be found online[1].

The experimental setup is depicted in Fig. 4(a). The purpose of the test is to overcome the depicted markers with the end-effector, drawing a square. Marker placement was chosen to stress the robot configurations to be close to joint limits (see Fig. 4(b)) and kinematic singularity (see Fig. 4c). The tests were performed using the Kuka LWR iiwa robot, controlled via ROS middleware, running on GNU/Linux OS as from Hennersperger et al. (2016).

Regarding the practical implementation of dynamic task composition, each task is represented by a static matrix, while the related Jacobian is implemented as a dynamic matrix using C++ standard vector class[2] to prevent any issue related to memory reallocation when the dimensions of the Jacobian matrix changes with respect to the activated tasks.

The following gains regarding the proposed methodology were experimentally tuned: the gain for the force task is equal to $\boldsymbol{\Lambda}_f = 0.02\boldsymbol{I}_3$; the gain for the

[1] https://youtu.be/-ZHwauD0NQY.

[2] https://en.cppreference.com/w/cpp/container/vector.

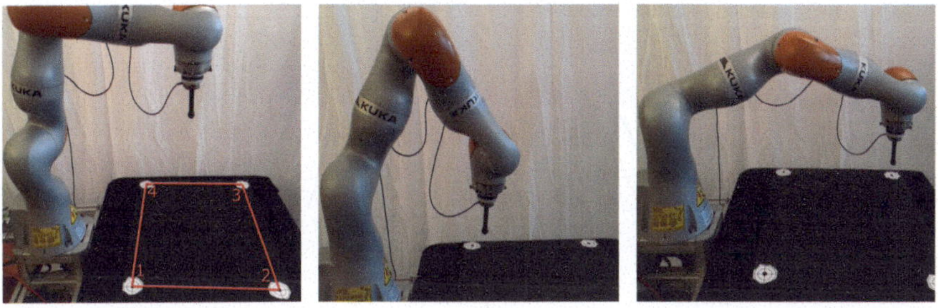

(a) (b) (c)

Fig. 4. Experimental setup: (a) the operator must reach the visual markers with the end-effector of the robot, bringing the robot configuration close to the mechanical joint limits (b) and to a kinematic singularity (c).

position task is equal to $\boldsymbol{\Lambda}_p = 5.5\boldsymbol{I}_3$; the gain for the orientation task is equal to $\boldsymbol{\Lambda}_o = 5.5\boldsymbol{I}_3$; the gain for the manipulability task is equal to $\boldsymbol{\Lambda}_m = 0.1\boldsymbol{I}_3$; the gain for the joint limits task is equal to $\boldsymbol{\Lambda}_l = 0.001\boldsymbol{I}_3$; the threshold to activate the manipulability stack is set to $\bar{\sigma}_m = 1.5$; the threshold to activate the joint limit stack is set to $\bar{\sigma}_l = 2$ rad. The general admittance control scheme is instead taken from Siciliano et al. (2009), with the following values for the apparent dynamics of the end-effector position: desired mass matrix $\boldsymbol{I}_3$; desired damping matrix $100\boldsymbol{I}_3$; desired stiffness matrix $0.001\boldsymbol{I}_3$.

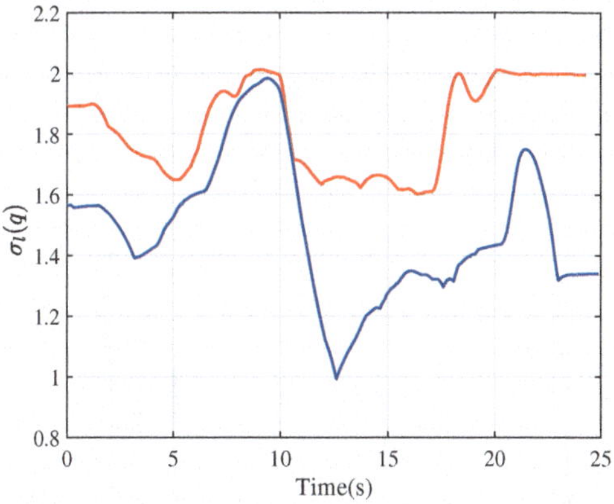

Fig. 5. Time history of the $\sigma_l(\boldsymbol{q})$ index. In blue, during the proposed approach test. In red, during the admittance control test.

A human operator performed the first test with the proposed methodology and, immediately after, one with an admittance control scheme. At any level of precision, it was impossible for the human operator to replicate the same trajectories exactly. Moreover, since the two experiments were performed in cascade,

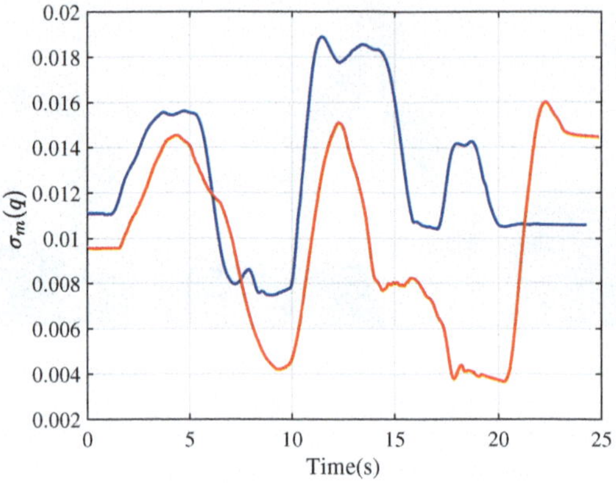

Fig. 6. Time history of the $\sigma_m(\boldsymbol{q})$ index. In blue, during the proposed approach test. In red, during the admittance control test.

the initial conditions of the two tests were slightly different as well. The time history of the $\sigma_l(\boldsymbol{q})$ index is depicted in Fig. 5. In blue, the one obtained through the proposed approach, in red the one obtained through the admittance control scheme. It is clear that the proposed methodology outperforms the classical admittance control scheme since the joint limit avoidance task is explicitly taken into account. The history of the stack switching in Fig. 7 highlights that the procedure made suitable changes in the task priority to deal with those situations in which some joint was very far from its desired value. The time history of the $\sigma_m(\boldsymbol{q})$ index is depicted in Fig. 6. In blue, the one obtained through the proposed approach, in red the one obtained through the admittance control scheme. It is again clear that the proposed methodology outperforms the classical admittance control scheme for most of the experiment. At the end of the test, the admittance

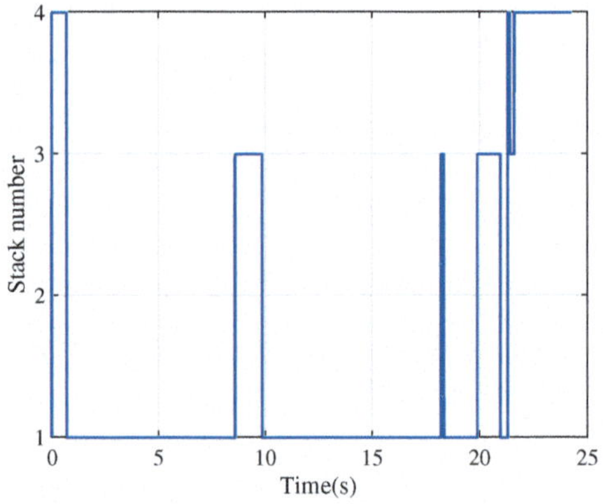

Fig. 7. Time history of the switching between the stack during the proposed approach test. Stack 1 is the force stack. Stack 2 is the manipulability stack. Stack 3 is the joint limit stack. Stack 4 is the pose stack.

control obtained a better configuration regarding the robot manipulability. This can be anyway handled by the proposed methodology by adding, for instance, the manipulability as a lower priority task within the pose stack.

7 Conclusion and Future Work

A hierarchical prioritized task control with dynamic task composition was presented in this paper to allow an effective co-manipulation for redundant manipulators. In the proposed approach, the kinematic redundancy of the robot was exploited to increase the performance of the manipulator during human-robot physical interaction. In particular, a set of primary and secondary tasks were composed to avoid mechanical joint limits and kinematic singularity during the execution of a shared task. The human operator was allowed to freely move the position of the robot end-effector through a proper force sensor mounted on the tip of the manipulator. To test the benefits of the proposed cobot system, a standard admittance control was compared to the proposed approach. The collected results demonstrated the effectiveness of the system.

Future work is dedicated to adding a phase in which the movements of the human operator can be employed to predict which of the secondary tasks is going to fail and re-arrange the prioritized stack accordingly.

Acknowledgments. The research leading to these results has been partially supported by the H2020-ICT-731590 REFILLS project; the WELDON project, in the frame of Programme STAR, financially supported by UniNA and Compagnia di San Paolo; the PRINBOT project, in the frame of the PRIN 2017 research program with grant number 20172HHNK5_002. The authors are solely responsible for its content.

References

Aertbeliën, E., De Schutter, J.: eTaSL/eTC: a constraint-based task specification language and robot controller using expression graphs. In: 2014 IEEE/RSJ International Conference on Intelligent Robots and Systems, pp. 1540–1546 (2014)

Agravante, D., Cherubini, A., Bussy, A., Gergondet, P., Kheddar, A.: Collaborative human-humanoid carrying using vision and haptic sensing. In: 2014 IEEE International Conference on Robotics and Automation, pp. 607–612 (2014)

Antonelli, G.: Stability analysis for prioritized closed-loop inverse kinematic algorithms for redundant robotic systems. IEEE Trans. Robot. **25**, 985–994 (2009)

Baerlocher, P., Boulic, R.: Task-priority formulations for the kinematic control of highly redundant articulated structures. In: 1998 IEEE/RSJ International Conference on Intelligent Robots and Systems, pp. 323–329 (1998)

Buonocore, L., Cacace, J., Lippiello, V.: Hybrid visual servoing for aerial grasping with hierarchical task-priority control. In: 2015 23rd Mediterranean Conference on Control and Automation, pp. 617–623 (2015)

Caccavale, F., Lippiello, V., Muscio, G., Pierri, F., Ruggiero, F., Villani, L.: Grasp planning and parallel control of a redundant dual-arm/hand manipulation system. Robotica **31**(7), 1169–1194 (2013)

Caccavale, R., Cacace, J., Fiore, M., Alami, R., Finzi, F.: Attentional supervision of human-robot collaborative plans. In: 2016 25th IEEE International Symposium on Robot and Human Interactive Communication, pp. 867–873 (2016)

Casalino, G., Zereik, E., Simetti, E., Torelli, S., Sperindé, A., Turetta, A.: A task subsystem priority based control strategy for underwater floating manipulators. IFAC Proc. Vol. **45**(5), 170–177 (2012)

Chiaverini, S.: Singularity-robust task-priority redundancy resolution for real-time kinematic control of robot manipulators. IEEE Trans. Robot. Autom. **13**(3), 398–410 (1997)

Corrales, J.A., Garcia Gomez, G.J., Torres, F., Perdereau, V.: Cooperative tasks between humans and robots in industrial environments. Int. J. Adv. Robot. Syst. **9**, 94 (2012)

Dimeas, F., Moulianitis, V., Aspragathos, N.: Manipulator performance constraints in human-robot cooperation. Robot. Comput.-Integr. Manuf. **50**, 222–233 (2018)

Dubey, R., Luh, J.Y.S.: Redundant robot control using task based performance measures. J. Field Robot. **5**, 409–432 (1988)

Ficuciello, F., Romano, A., Villani, L., Siciliano, B.: Cartesian impedance control of redundant manipulators for human-robot co-manipulation. In: 2014 IEEE/RSJ International Conference on Intelligent Robots and Systems, pp. 2120–2125 (2014)

Gullapalli, V., Grupen, R., Barto, A.: Learning reactive admittance control. In: 1992 IEEE International Conference on Robotics and Automation, pp. 1475–1480 (1992)

Hennersperger, C., Fuerst, B., Virga, S., Zettinig, O., Frisch, B., Neff, T., Navab, N.: Towards MRI-based autonomous robotic US acquisitions: a first feasibility study. IEEE Trans. Med. Imaging **36**(2), 538–548 (2016)

Labrecque, P., Gosselin, C.: Variable admittance for pHRI: from intuitive unilateral interaction to optimal bilateral force amplification. Robot. Comput.-Integr. Manuf. **52**, 1–8 (2018)

Landi, C., Ferraguti, F., Sabattini, L., Secchi, C., Fantuzzi, C.: Admittance control parameter adaptation for physical human-robot interaction. In: 2017 IEEE International Conference on Robotics and Automation, pp. 2911–2916 (2017)

Lippiello, V., Cacace, J., Santamaria-Navarro, A., Andrade-Cetto, J., Trujillo, M., Esteves, Y., Viguria, A.: Hybrid visual servoing with hierarchical task composition for aerial manipulation. IEEE Robot. Autom. Lett. **1**(1), 259–266 (2016)

Mansard, N., Stasse, O., Evrard, P., Kheddar, A.: A versatile generalized inverted kinematics implementation for collaborative working humanoid robots: the stack of tasks. In: 2009 International Conference on Advanced Robotics (2009)

Nakamura, Y., Hanafusa, H., Yoshikawa, T.: Task-priority based redundancy control of robot manipulators. Int. J. Robot. Res. **6**(2), 3–15 (1987)

Newman, W.S.: Stability and Performance limits of interaction controllers. J. Dyn. Syst. Meas. Control **114**(4), 563–570 (1992)

Rocchi, A., Hoffman, E., Caldwell, D., Tsagarakis, N.: OpenSoT: a whole-body control library for the compliant humanoid robot COMAN. In: 2015 IEEE International Conference on Robotics and Automation, pp. 6248–6253 (2015)

Siciliano, B., Sciavicco, L., Villani, L., Oriolo, G.: Robotics: Modelling, Planning and Control. Springer, London (2009)

Yoshikawa, T.: Manipulability of robotic mechanisms. Int. J. Robot. Res. **4**(2), 3–9 (1985)

Zanchettin, A., Rocco, P.: Probabilistic inference of human arm reaching target for effective human-robot collaboration. In: 2017 IEEE/RSJ International Conference on Intelligent Robots and Systems, pp. 6595–6600 (2017)

How to 3D-Print Compliant Joints with a Selected Stiffness for Cooperative Underactuated Soft Grippers

Irfan Hussain[1], Zubair Iqbal[2], Monica Malvezzi[2,3], Domenico Prattichizzo[2,3], and Gionata Salvietti[2,3](✉)

[1] Khalifa University Center for Autonomous Robotic Systems (KUCARS), Khalifa University of Science and Technology, Abu Dhabi, UAE
irfan.hussain@ku.ac.ae

[2] Dipartimento di Ingegneria dell'Informazione, Università degli Studi di Siena, Via Roma 56, 53100 Siena, Italy
salviettigio@diism.unisi.it

[3] Department of Advanced Robotics, Istituto Italiano di Tecnologia, Via Morego 30, 16163 Genoa, Italy

Abstract. Soft robotics is an expanding area of research which exploits compliance and adaptability of soft structures to design highly adaptive robots for contact interactions. The ability to 3D-print materials with softer, more elastic materials properties is a recent development and a key enabling technology for the rapid development of soft robots. Among the possible applications, soft cooperative grippers and manipulators have demonstrated a high potential impact. However, retrieving data about mechanical properties of such novel soft materials and information about the parameters to select in the 3D-printers is often not straightforward. The aim of this chapter is to systematically investigate the mechanical properties of 3D-printed specimens from one of the most used soft filament, the Ninjaflex (Lulzbot, USA). In particular, we focus on the characterization of bending stiffness with reference to the selection of different infill density and printing patterns. We also report on how the collected data can be used in the design phase to obtain a desired behaviour of a soft gripper and on repeatability of the 3D-printing process. Finally, we tested with a finger of a cooperative gripper how the different approaches to obtain a given stiffness value affect the flexion/extension trajectory. This work is intended to be a useful tool for researchers who use rapid-prototyping techniques to develop soft grippers and it also serves as a possible guideline to the characterization of the mechanical properties of novel soft filaments.

Keywords: Soft grippers · Rapid prototyping

1 Introduction

Soft grippers are an interesting and promising direction of research in robotic hands leading toward solutions that are characterized by a simplified hardware

F. Ferraguti et al. (Eds.): HFR 2019, SPAR 12, pp. 139–153, 2020.
https://doi.org/10.1007/978-3-030-42026-0_11

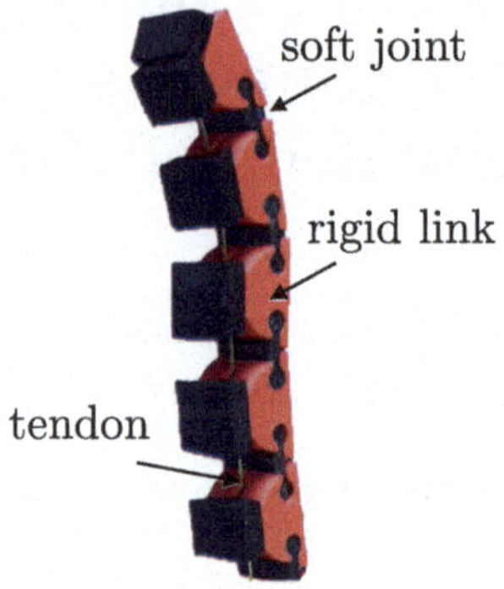

Fig. 1. Example of the structure of a finger of the CoGripper [6]. Soft joints connect rigid links and the finger is actuated by a tendon.

structure, with a limited number of actuators, where the needed adaptability is guaranteed by the presence of passive compliant elements [1,2] or by the compliant structure of the hand [3]. Exploiting compliance and underactuation in robotic hands enable to improve the adaptability of the devices to unstructured environments and the capability of manipulating fragile objects with uncertain shapes. Interesting applications are the harvesting or the manipulation of fruit and vegetables [4]. In such devices, a proper design of passive elements is paramount to guarantee the suitable contact force distribution [5]. Collaborative grippers are another possible application, where compliance can guarantee a safe interaction with the human operator [6].

The design process of soft gripper differs from that of traditional robotic structures. Rigid robots are composed of rigid links connected to actuators and, in such systems, the problems of structural behaviour and device control are basically decoupled. On the other hand, in the emerging field of soft robotics, these boundaries are blurred and there is a need to define new paradigms in engineering that challenged us to re-examine the relationships between materials and mechanisms behaviour and device performance and control [7]. In particular, the advent of the so-called 3D-printing technology changed modern manufacturing [8]. 3D-printed parts based on the fused deposition method (FDM) are widely used in many different applications including robot design and fabrication [9]. New materials introduced by additive manufacturing techniques opens a set of possibilities for designers [10]. Beside geometric shape, additional parameters such as chemical composition, surface treatments, infill density percentage, printing pattern, etc. can be used to get the desired mechanical properties [11]. In underactuated compliant robots, the lack of actuation can be partially compensated by regulating, during the design phase, the stiffness and deformation of compliant elements.

In [12,13], we presented our preliminary studies on modular underactuated robotic fingers, where we exploited the possibility of changing passive joint

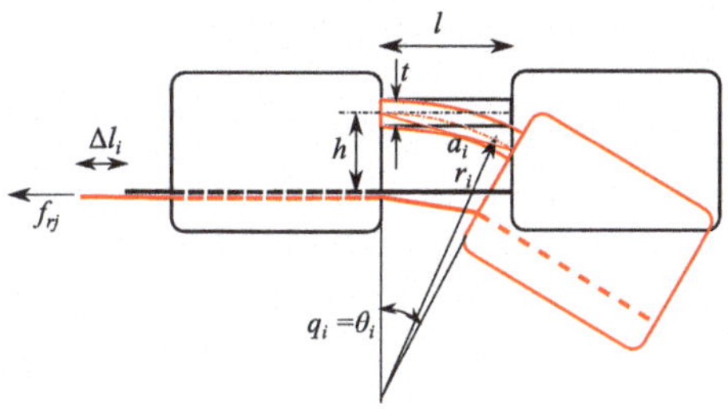

Fig. 2. Main geometric parameters of the passive joint described in [13].

stiffness to obtain a desired trajectory for the fingertip. The finger structure studied is reported in Fig. 1. We adopted a commercially available thermoplastic elastometer (TPE), called Ninjaflex [14]. TPEs are materials that has the ability to undergo large deformations before failure and present good mechanical properties [15]. In the above mentioned works, we evaluated how different percentage of infill density affected the bending stiffness of the soft joints. In [16], a set of experimental tests were performed to evaluate the sensitivity of component hardness and tensile strength with respect to infill density percentage. Apart from percentage infill density there are other key parameters which also affect the mechanical properties of 3D-printed components, e.g. printing patterns, which need to be exploited.

The lack of easily accessible and organized information about the printing procedures and the mechanical properties of soft filament motivated the research presented in this chapter. Our main contribution is the systematic analysis of the bending stiffness of 3D-printed samples obtained using different settings of the 3D-printer. We tested and provide guidelines on how to design the stiffness by changing the infill density and the printing pattern. These parameters are common for all the 3D-printers which support soft filaments. We also tested how geometric properties of the specimen affect the stiffness and in particular we tested samples with different length. Finally, we tested how properties are accurately reproducible among samples printed with the same settings. Although we tested only one type of soft filament, we do believe that the procedures reported in this work can be easily reproduced to also test other materials. Studies on the stiffness regulation are mainly motivated by the design of soft finger like that reported in Fig. 1. However, the characterization of the stiffness could also be exploited for the fabrication of fully soft 3D-printed structures similar to that presented in [17].

The rest of the chapter is organized as follows. In Sect. 2, we describe the material properties and the methods used to test the specimens. In Sect. 3, we report the results obtained as well as a case study. Finally, in Sect. 4 conclusion and future work are outlined.

2 Characterization of the Stiffness

Let us consider a passive joint connecting two links of an underactuated tendon-driven finger as the one described in [13], schematically shown in Fig. 2. Let us indicate with i and j the indexes corresponding to the joint and tendon, respectively. In a passive joint, we can approximately relate joint angular deflection θ_i to the applied torque τ_i by the simplified linear relationship

$$\tau_i = k\theta_i, \tag{1}$$

where k indicates joint flexural rigidity or bending stiffness. In [13], we discussed about the limitations of such an approximation, for a complete description the reader can refer for instance to [18]. The stiffness k depends mainly on joint geometry and material structural properties, and in particular, considering a linear elastic behaviour, on Young's modulus E, i.e. $k = k(\mathbf{d}, E)$, where $\mathbf{d}$ is a vector containing all the parameters defining joint geometry (e.g., for a parallelepiped joint, its length l, width w and thickness t, see Fig. 2). Young's modulus E depends on material parameters and fabrication methods, i.e. $E = E(p_1, p_2, \cdots, p_n)$, where each value p_i indicates one specific material property. In this chapter, we investigate the dependency of E with respect to two specific parameters

$$E = E(p_1, p_2), \tag{2}$$

where p_1, indicates the printing pattern and p_2, indicates the percentage infill density. Similarly, we can express the overall joint stiffness measured in the tests proposed in the following as

$$k = k(\mathbf{d}, p_1, p_2). \tag{3}$$

The main objective is to experimentally define passive joint stiffness properties, and in particular the relationship between joint deflection and applied torque, as a function of the two printing parameters. Although the general principles for determining the tensile properties of the materials under defined conditions are described in specific regulations [19], we designed a customized test, good enough to obtain directly the data necessary to design the passive joints in modular underactuated hands. In our tests, we considered as specimens 3D-printed elements resembling the soft joints (see Fig. 1) of the robotic finger described in [13] and summarized in Sect. 3.2. In particular, the specimen used is a flexible hinge acting as joint between two rigid links. The dimensions of the specimen are $24 \times 20 \times 3.75$ mm.

2.1 Material and Printer Settings

The soft filament considered in this work is a commercially available thermoplastic polyurethane (TPU) named NinjaFlex (Lulzbot, USA). NinjaFlex is one of the most widely used material thanks to its high flexibility and longevity compared to non-polyurethane materials [14]. Its consistency in diameter and

Table 1. Mechanical and thermal properties of Ninja flex material.

Mechanical properties	
Tensile strength, yield	4 Mpa
Tensile strength, ultimate	26 Mpa
Tensile modulus	12 Mpa
Elongation at yield	65%
Elongation at break	660%
Hardness	85 Shore A
Impact strength	4.2 $\frac{KJ}{m^2}$
Abrasion resistance (mass loss, 10,000 cycles)	0.08 g
Thermal properties	
Melting point	216 °C
Glass transition (Tg)	−35 °C
Heat Deflection Temperature (HDT) @ 0.07 MPa	60 °C

roundness is competitive with respect to other polyurethane materials. It is made of a specially formulated TPU material which possesses a low-tack, easy-to-feed texture. The result are very flexible and resistant prints ideal for direct-drive extruders. The mechanical and thermal properties of the NinjaFlex material are summarized in Table 1.

Most of the available 3D-printers are equipped with spring loaded feeder that are capable of extruding NinjaFlex filament. We used the Lulzbot Taz 5 3D-printer [20].

We used the software tool "Slic3r" [21] to translate the digital 3D models into instructions for the 3D-printer. These instructions are in the form of G Codes, which are basically coordinates in the form of X, Y and Z axis commanding the motor where to move, how fast to move and which path to follow. Slic3r segments the model into layers and generates a path to fill them (see, Fig. 3). 3D models can be generated using any Mechanical Computer Aided Design (MCAD) software and is transformed in STL (STereoLithography), OBJ (Object) or AMF (Additive Manufacturing File) format. This format contains layer by layer information of the part. After the import, Slic3r gives an option to orient the part during the printing phase. It is worth to recall that the orientation of the part plays a very important role in the realization of the desired printing pattern and on the overall mechanical properties of the component [22]. In particular, for the flexible joints described in this chapter, keeping the part flat on the bed along its length allows us to realize the complete pattern. Once the model is oriented, the software tool allows us to set different options, e.g. choosing the extruder to use, the printing pattern, the value of percentage infill density, the top and bottom layer pattern, etc. Once these parameters are set, the G Code can be generated and fed to the printer.

2.2 Test Bench Setup

Slic3r supports seven possible infill patterns. There are several considerations to take into account when selecting the more suitable infill pattern, e.g., the required object strength, the material being used to print the parts, and the time required to print the part. Note that the time taken to print the part and the material consumed depends on the geometry of the part, i.e., if we select a complex pattern it will take more time to print and more material will be used.

In this paper, we considered all the seven different patterns to print the flexible hinge part, as shown in Fig. 3. Therefore, variable p_1 in Eqs. (2) and (3) can assume seven values:

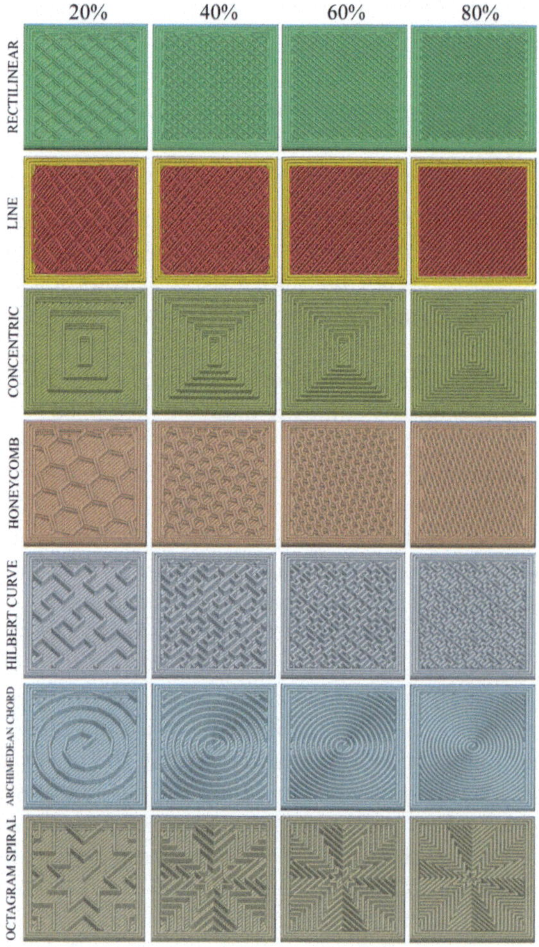

Fig. 3. Infill patterns at varying densities generated in Slic3r. Left to right: 20%, 40%, 60%, 80%. Top to bottom: Rectilinear, Line, Concentric, Honeycome, Hilbert Curve, Archimedean Chords and Octagram Spiral.

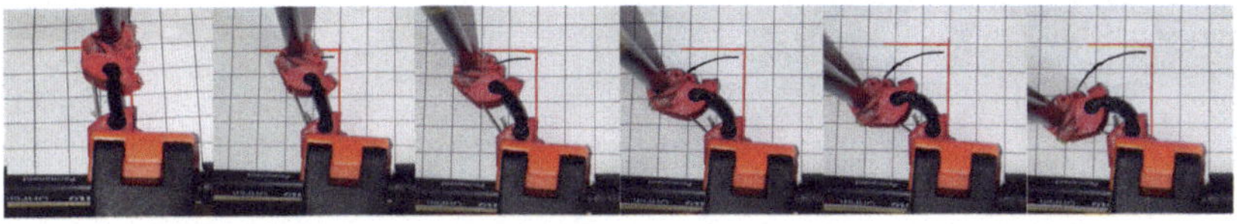

Fig. 4. Experimental setup for recording the bending angle of the tested specimen. The two rigid links are connected through the sample that acts as a single flexible joint. The bending trajectory of flexible part in response to the applied force is recorded using a marker connected to the tip of the mobile part.

- $p_{1,1}$: Rectilinear;
- $p_{1,2}$: Line;
- $p_{1,3}$: Concentric;
- $p_{1,4}$: Honeycomb;
- $p_{1,5}$: Hilbert Curve;
- $p_{1,6}$: Archimedean Chords;
- $p_{1,7}$: Octagram Spirals.

Another important setting to be taken into account is the infill density, that can be defined as the amount of material deposited by the printer. The higher the infill density is, the stiffer the printed part is expected to be. Infill density is typically expressed as a percentage and can be varied from 10% to 100%, where 10% is the lowest infill density resulting in a very flexible/soft output part, while 100% infill density will produce a stiffer and very tightly packed model, which needs more material. We considered four different values of infill density to analyse its effect on passive joint's overall stiffness, therefore variable p_2 in Eqs. (2) and (3) assumes four possible values in our tests:

- $p_{2,1}$: 20%;
- $p_{2,2}$: 40%;
- $p_{2,3}$: 60%;
- $p_{2,4}$: 80%.

The third important setting that we considered while printing a model is the top and bottom infill pattern, because it holds the complete structure together providing top and bottom cover to the infill pattern. The top and bottom layers contains the infill layer in which we varied parameters p_1 and p_2 parameters. There are five different patterns available: (*i*) Rectilinear, (*ii*) Concentric, (*iii*) Hilbert Curve, (*iv*) Archimedean Chords, (*v*) Octagram Spiral. In the specimens considered, the top and bottom layers are printed at infill density of 100%, with Rectilinear pattern.

Finally, another possible way to achieve different stiffness values is through the variation in the geometry of the part. In this paper, we considered, as an example, the variation of the sample length. We realized four different geometries with length

- $l_1 = 24$ mm;

- $l_2 = 30$ mm;
- $l_3 = 36$ mm;
- $l_4 = 42$ mm;

Printing settings are specific for the printer being used. We selected the following settings in case of the Lulzbot printer (TAZ 5), being used in this work:

1. no retract;
2. uniform print speed: 15 mm/s;
3. multi-loop skirt: 4 loops;
4. hot-end temperature: 240 °C;
5. print bed temperature: 40 °C;
6. travel speed: 100 mm/s;
7. extrusion width: 0.5 mm with a 0.5 mm nozzle;
8. layer height: 0.4 mm.

The experimental setup to measure the bending profile of the generated flexible specimens is shown in Fig. 4. We considered a single module of a robotic finger of the gripper introduced in [6]. Two rigid parts are connected with a single flexible joint. One of the rigid parts is grounded (fixed), whereas the second rigid part is free to move parallel to a fixed plane. A marker is attached at the free tip of the rigid part, that is bent by applying a known force to the tendon. In the initial unloaded configuration the joint is fully open and the passive element is fully straight. Starting from this configuration, a force is being applied to bend the flexible part. The applied force is measured through a dynamometer (Vernier, USA). While the joint bends, the marker draws the resultant trajectory, as reported in Fig. 4. For each test, we recorded the trajectory produced by the joint deflection as a function of the applied force. We repeated this process for all the 28 specimens presented in Fig. 3. By assuming that the trajectory is a circular arc, the corresponding deflection angle, necessary to estimate joint stiffness according to the relationship in Eq. (1), can be evaluated as

$$\theta_i = \frac{a_i}{\rho_i} \tag{4}$$

where a_i indicates the trajectory arc length, and ρ_i the trajectory curvature radius.

3 Results

3.1 Test Bench Results

We printed all the samples and measured the bending of each part in response to the applied force through the experimental setup shown in Fig. 4. For each printing pattern, four values of the percentage infill density have been considered.

We measured the bending of each part in response to the applied force through the proposed experimental setup, the obtained results are shown in Fig. 5. As expected, for all the examined patterns the 20% infill density produced the softest samples, whereas the 80% the stiffest one. Note that the samples with Line and Archimedean Chords patterns at infill density of 60% and 80% the behaviours of the samples are very similar, with a very narrow difference in the deflection. So it can be concluded that at 60% infill density the sample will reach its stiffest structure. Increasing the infill density beyond this value will not change significantly the behaviour of the sample and will result in a consumption of material and time. Interestingly, some patterns resulted to be less sensitive to infill density as compared to others, see e.g. the Archimedean Chords, whereas the Rectilinear and Line patterns present the highest difference in bending angle in response to the same applied force.

In Fig. 6, we report joint stiffness as a function of the applied force. We consider the same 28 configurations previously introduced, in this case we reported the values obtained with different printing patterns for the same infill density percentage. Ideally, the stiffness should be constant while applying an increasing force. Interestingly, this behaviour is not respected for all the applied forces. Stiffness variation depends on the infill density percentage: samples printed with 20% and 40% of infill density percentage (Fig. 6a and b) lead to a stiffness value that is less sensitive on the applied forces, while in samples printed with 60% and 80% (Fig. 6c and d) present a stiffness with an higher variability. Also the printing patterns influence the stiffness behaviour, especially for high infill density, see Fig. 6d.

Figure 7 shows the results of variation of a geometric parameter, in particular the length of the sample l. We printed four different specimens with a length of $24, 30, 36$, and 42 mm. All the samples were obtained selecting an infill density of 60% and Honeycomb pattern. We can notice that, as expected, the length and the stiffness are inversely proportional. Parts tends to move from stiff to flexible structure upon increasing the length.

Finally, we evaluated the repeatability of the mechanical properties of the samples considering how accurately a 3D-printer can print same samples. To observe the repeatability of the printing process, it was necessary to print a number of sample with identical settings and then check their behaviour. Figure 8 shows the results obtained with 5 different samples printed with a Hilbert Curve pattern and an infill density of 60%. We tested each sample by the same experimental setup. Each specimen was subjected to 10 different force values and the bending angle was measured. The results shown in Fig. 8 show that the printed samples have a quite similar behaviour. It can be concluded that the printer has a repeatability performance suitable for our applications, since each batch of parts printed has similar characteristics.

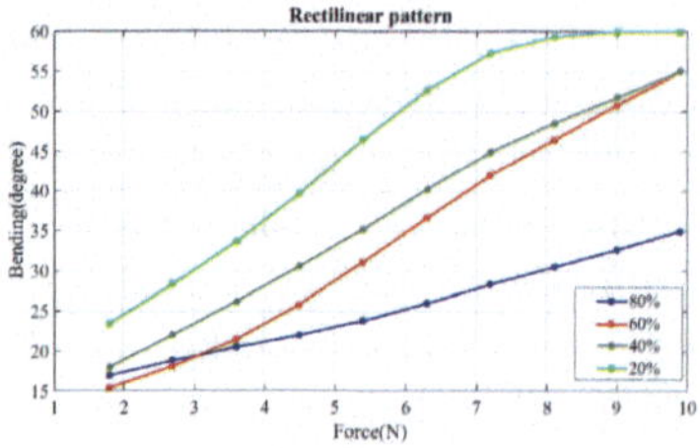

(a) Rectilinear.

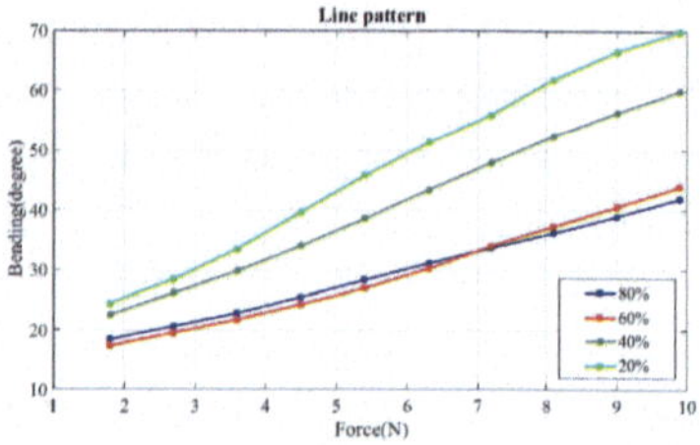

(b) Line.

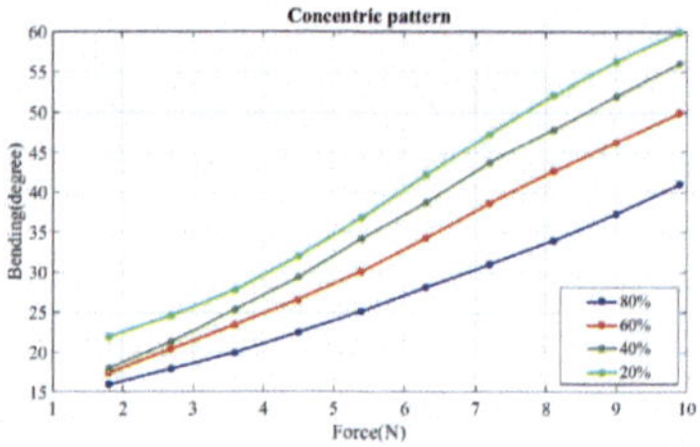

(c) Concentric.

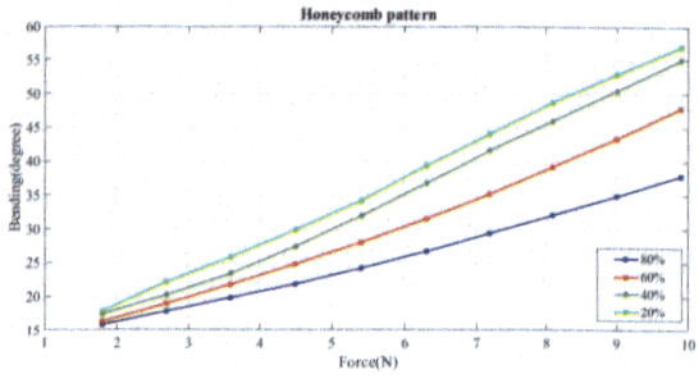

(d) Honey Comb.

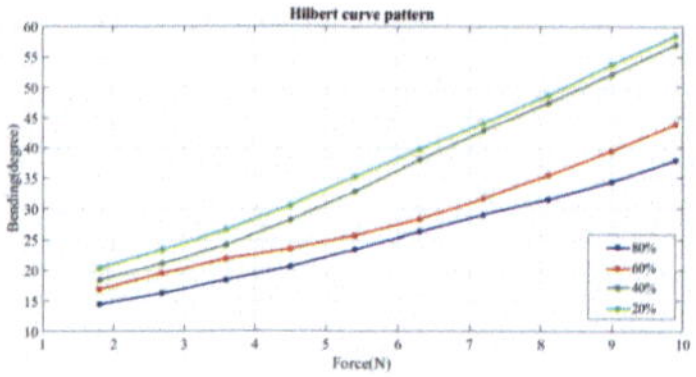

(e) Hilbert Curve.

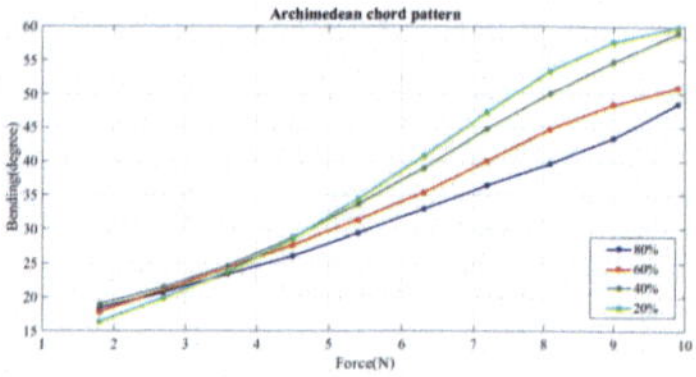

(f) Archimedean Chords.

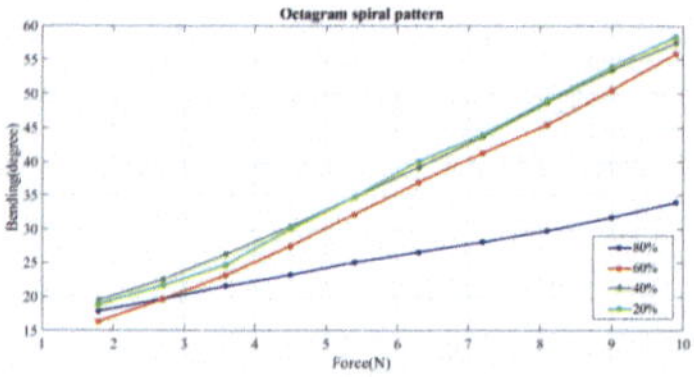

(g) Octagram.

Fig. 5. Sample rotations as a function of the applied force, varying infill density percentage for each pattern. Results shows that the pattern influences both the curve behaviour and the influence of infill density percentage: in some patterns, as for instance the Rectilinear one, the weight of infill density percentage is relevant, while in others, as for instance the Archimedean chord, the curves obtained with different infill density percentages are more similar.

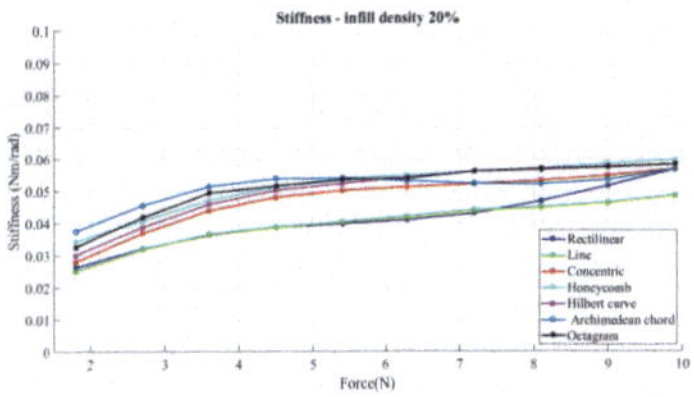

(a) 20% infill density

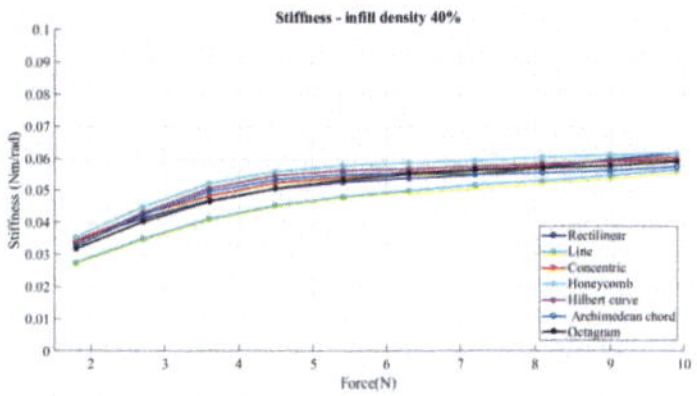

(b) 40% infill density.

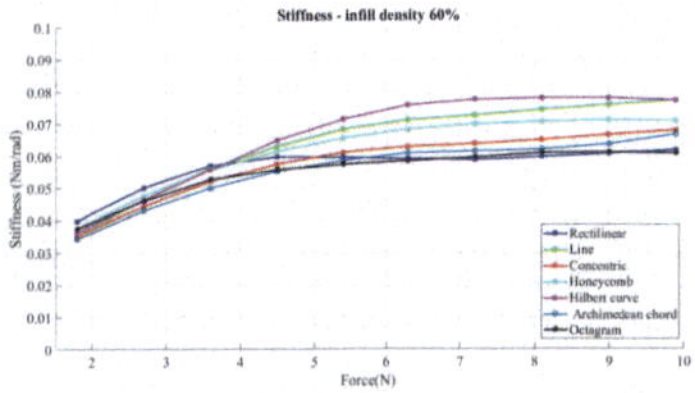

(c) 60% infill density

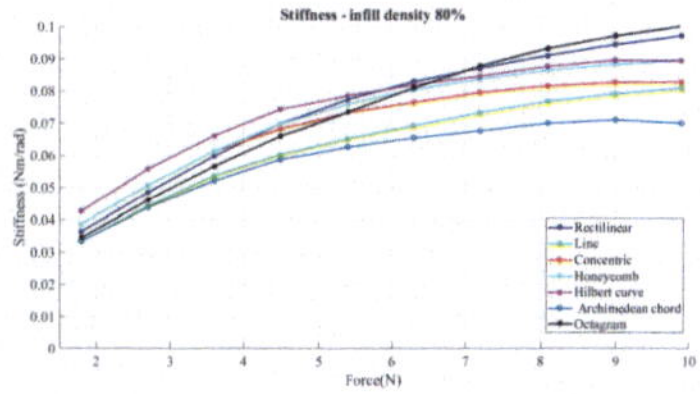

(d) 80% infill density

Fig. 6. Stiffness variation due to different pattern types while keeping the same infill density

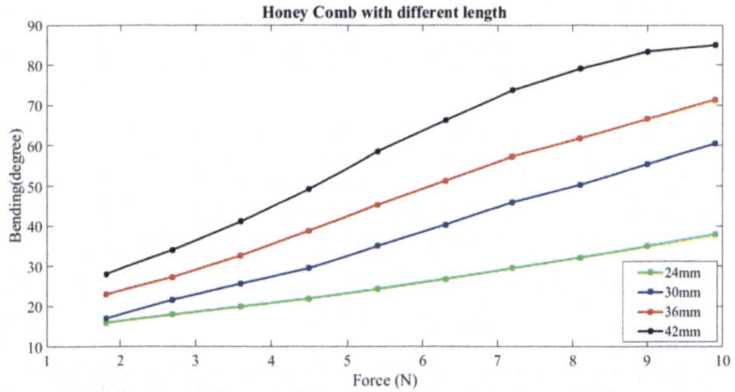

Fig. 7. Variation in joints bending due to different length of flexible joint.

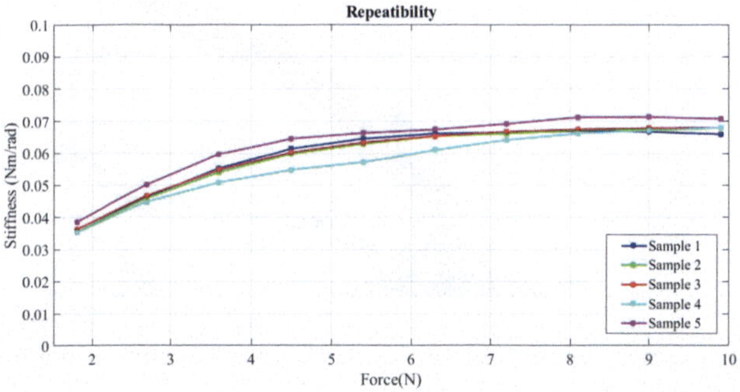

Fig. 8. Five samples of Hilbert Curve having in fill density of 60%. Each sample was subjected to 10 different Force values and the corresponding deflection angle was measured.

3.2 Case Study

In this subsection, we propose as a case study the design of passive joints for a finger of the CoGripper [6], a robotic device for cooperative manipulation tasks with humans. The gripper is composed of two pairs of fingers, actuated with two motors, that can be controlled in a coordinated way or independently. Each finger has a modular underactuated structure, composed of three phalanges connected by passive joints. The gripper is wireless, so it can be easily connected both to the robotic arms and on passive structures. The soft design of the robotic finger is shown in Fig. 1. The proposed device is intrinsically-compliant, modular, underactuated, and cable driven as shown in Fig. 9. We have demonstrated in [12, 13], that the fingertip trajectories of the device play an important role during the grasping phase. Moreover, we have shown that different fingertip trajectories can be achieved designing the stiffness ratio between two consequent soft joints. In these preliminary works, we have only used the variation of infill density to realize soft joints with a desired stiffness. Results obtained in this paper and discussed in the previous section allow us to integrate the design of passive joints by considering the possible combinations of infill density percentage and printing pattern. In particular in the following, we will compare two different solutions to select the same stiffness distribution allowing to realise a given trajectory for the fingertip by combining different infill density values and different printing pattern shapes.

We assembled two finger configurations by the stiffness of each joint, defined to obtain the desired trajectory according to the procedure described in [12,13], by selecting either a different infill density percentage with the same pattern, or a different pattern with the same infill density percentage. The aim of the experiment was to verify if it is possible to achieve the same fingertip trajectory

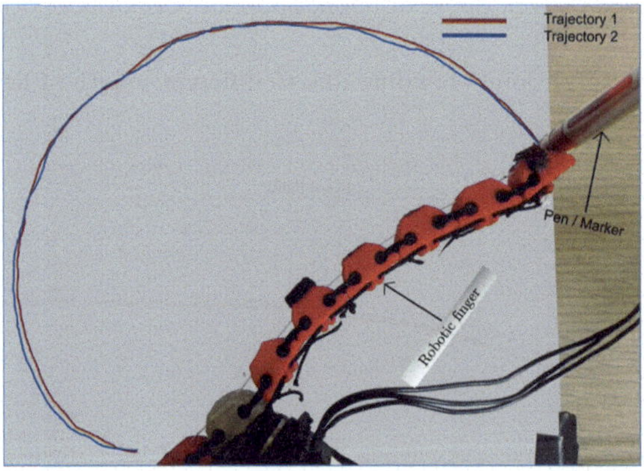

Fig. 9. The finger trajectory plotted by setting the similar joints stiffness through both percentage infill density and pattern variation. Blue trajectory is plotted with variation of percentage of infill density while the red one is achieved through the variation of pattern.

Table 2. Printing pattern and infill density for two different solutions to select the desired joints stiffness of the finger to two trajectories shown in Fig. 9.

Stiffness values (Nm/rad)	Pattern type @$p_{2,3}$: 60% (Trajectory 1)	Percentage of infill density @$p_{1,1}$: Rectilinear (Trajectory 2)
0.06	Octagram Spirals	20
0.064	Rectilinear	32
0.068	Archimedean Chords	44
0.072	Concentric	56
0.076	Honeycomb	68
0.08	Hilbert Curve	80

by setting the joints stiffness through both methods. The experimental setup is reported in Fig. 9. Table 2 shows the resultant printing pattern and infill densities for both solutions to select desired joints stiffness of finger used in our experiment. The support base of the finger was fixed on the table through a clamp and a marker was attached at the fingertip. A white paper was fixed at the surface of the table so that the device could draw the resultant trajectory. The robotic finger was moved through its actuator from fully open position (extended) to completely bent (flexed). Figure 9 shows the snapshots of the two resultant trajectories. As it can be seen, the trajectory are quite similar, not withstanding all the uncertainties and approximations that this type of passive and underactuated devices inherently present. Both the infill density percentage and the pattern profile can thus be effectively exploited to achieve a desired joint stiffness.

4 Conclusion

New additive manufacturing technologies and materials open a completely new set of design opportunities. In particular, exploiting them in underactuated compliant mechanical structures can be a strategic solution for the design of simple, highly adaptable and robust robotic hands that can be effectively employed in industrial and service applications [6,23]. 3D printing technology has tremendous potential in many engineering applications. However, to fully utilize this potential, there is the need to investigate and analyse the different printing possibilities and usage of tools effectively. In this chapter, we present a systematic analysis of the stiffness of several samples obtained with different 3D-printer settings. We have reported how different infill density and printing pattern affect the mechanical behaviour of specimens obtained using a soft filament called NinjaFlex. We believe that this study will contribute to get systematic way to design and realize the desired stiffness of flexible elements of a structure through 3D printing and hence optimizing the development in terms of cost and time. Moreover, such guidelines have the tendency to fill the gap that the developers experience during the development of passive elements of soft grippers. The proposed method to analyse the stiffness could be easily extended to other soft filaments and materials. Currently, we are investigating how the proposed results can be exploited for the realization of 3D-printed soft grippers entirely fabricated using soft filaments. We are also combining more fingers to realize more complicated soft grippers.

Acknowledgement. The research leading to these results has received funding from the EU Horizon 2020/2015 project No 688857 "SOFTPRO".

References

1. Dollar, A.M., Howe, R.D.: The highly adaptive SDM hand: design and performance evaluation. Int. J. Robot. Res. **29**(5), 585–597 (2010)
2. Grioli, G., Catalano, M., Silvestro, E., Tono, S., Bicchi, A.: Adaptive synergies: an approach to the design of under-actuated robotic hands. In: 2012 IEEE/RSJ International Conference on Intelligent Robots and Systems (IROS), pp. 1251–1256. IEEE (2012)
3. Deimel, R., Brock, O.: A compliant hand based on a novel pneumatic actuator. In: 2013 IEEE International Conference on Robotics and Automation (ICRA), pp. 2047–2053. IEEE (2013)
4. Wang, Z., Torigoe, Y., Hirai, S.: A prestressed soft gripper: design, modeling, fabrication, and tests for food handling. IEEE Robot. Autom. Lett. **2**(4), 1909–1916 (2017)
5. Davidson, J.R., Mo, C.: Mechanical design and initial performance testing of an apple-picking end-effector. In: ASME International Mechanical Engineering Congress and Exposition (IMECE), pp. 13–19 (2015)
6. Salvietti, G., Iqbal, Z., Hussain, I., Prattichizzo, D., Malvezzi, M.: The co-gripper: a wireless cooperative gripper for safe human robot interaction. In: Proceedings of IEEE/RSJ International Conference on Intelligent Robots and Systems, Madrid, Spain, pp. 4576–4581 (2018)
7. Majidi, C.: Soft robotics: a perspective-current trends and prospects for the future. Soft Robot. **1**(1), 5–11 (2014)
8. Lipson, H., Kurman, M.: Fabricated: The New World of 3D Printing. Wiley, New York (2013)
9. Gao, W., Zhang, Y., Ramanujan, D., Ramani, K., Chen, Y., Williams, C.B., Wang, C.C., Shin, Y.C., Zhang, S., Zavattieri, P.D.: The status, challenges, and future of additive manufacturing in engineering. Comput. Aided Des. **69**, 65–89 (2015)
10. Truby, R.L., Lewis, J.A.: Printing soft matter in three dimensions. Nature **540**, 371 (2016)
11. Mutlu, R., Alici, G., in het Panhuis, M., Spinks, G.M.: 3D printed flexure hinges for soft monolithic prosthetic fingers. Soft Robot. **3**(3), 120–133 (2016)
12. Hussain, I., Salvietti, G., Malvezzi, M., Prattichizzo, D.: On the role of stiffness design for fingertip trajectories of underactuated modular soft hands. In: 2017 IEEE International Conference on Robotics and Automation (ICRA), pp. 3096–3101. IEEE (2017)
13. Salvietti, G., Hussain, I., Malvezzi, M., Prattichizzo, D.: Design of the passive joints of underactuated modular soft hands for fingertip trajectory tracking. IEEE Robot. Autom. Lett. **2**(4), 2008–2015 (2017)

14. Ninjaflex® the market leading flexible filament. https://ninjatek.com/products/filaments/ninjaflex/. Accessed 06 Sept 2017
15. Saari, M., Galla, M., Cox, B., Krueger, P., Cohen, A., Richer, E.: Additive manufacturing of soft and composite parts from thermoplastic elastomers. In: Solid Freeform Fabrication Symposium, pp. 949–958. University of Texas at Austin, Austin (2015)
16. Yarwindran, M., Azwani Sa'aban, N., Ibrahim, M., Periyasamy, R.: Thermoplastic elastomer infill pattern impact on mechanical properties 3d printed customized orthotic insole. ARPN J. Eng. Appl. Sci. **11**(10), 6519–6524 (2016)
17. Wang, Z., Chathuranga, D.S., Hirai, S.: 3d printed soft gripper for automatic lunch box packing. In: 2016 IEEE International Conference on Robotics and Biomimetics (ROBIO), pp. 503–508. IEEE (2016)
18. Odhner, L.U., Dollar, A.M.: The smooth curvature model: an efficient representation of Euler-Bernoulli flexures as robot joints. IEEE Trans. Robot. **28**(4), 761–772 (2012)
19. Plastics—Determination of tensile properties—Part 1: General principles. International Organization for Standardization, Geneva, March 2012
20. LulzBot TAZ 5. https://www.lulzbot.com/store/printers/lulzbot-taz-5. Accessed 06 Sept 2017
21. Slic3r, g-code generator for 3d printers. http://slic3r.org/. Accessed 06 Sept 2017
22. Chaudhry, S., Al-Dojayli, M., Czekanski, A.: Performance of 3-d printed thermoplastic polyurethane under quasi-static and high-strain rate loading. In: ASME 2016 International Mechanical Engineering Congress and Exposition, pp. V009T12A065–V009T12A065. American Society of Mechanical Engineers (2016)
23. Salvietti, G., Iqbal, Z., Malvezzi, M., Eslami, T., Prattichizzo, D.: Soft hands with embodied constraints: the soft ScoopGripper. In: Proceedings of IEEE International Conference on Robotics and Automation, Montreal, Canada, May 2019

Linking Human Factors to Assess Human Reliability

Fabio Fruggiero[1(✉)], Marcello Fera[2], Alfredo Lambiase[3], and Valentina Di Pasquale[3]

[1] School of Engineering, University of Basilicata, 85100 Potenza, Italy
fabio.fruggiero@unibas.it
[2] Department of Engineering, University of Campania "Luigi Vanvitelli", Aversa, CE, Italy
[3] Department of Industrial Engineering, University of Salerno, Fisciano, SA, Italy

Abstract. This study proposes a methodological review and an illustrative example for the Human Reliability Analysis. The paper deals with work-related stress and It discusses about the role of humans into manufacturing systems over automation control. A causal loop methodology is used to report the dynamic responses over time of humans at work. The Causal Loop is arranged inside a main framework which includes multi criteria decisions about main and subsidiaries concepts that influence the work-stress and the operations management. We chose the mapping method using expert, seniors and workers judgments and we demonstrated its utility in supporting decision for strategic operational goals. Factors and variables from the Human, Team, Work, Environment behavior are considered and declined in constant, measurable and depended elements with their inter and intra level connections. Contextualization in a safety based approach is reported. Findings are discussed with regard to a manufacturing set in terms of reliability management. We identified and modelled critical influencing factors for human reliability at work while discussing about human robotic interaction.

Keywords: Human Reliability · Human Factor · Work-stress · Safety · System Dynamics · Decision support systems

1 Introduction

Humans in manufacturing affect multiple goals and their behavior requires the analysis of interaction between different parts/elements/actors which constitute the inherent structure of the workplace. The role of humans as a strategic issue for the Operations Management (OM) is taken into consideration by the Human Factor (HF), which can be defined as "the scientific discipline that concerns with the understanding of the interaction between humans, and other elements in the system, in order to optimize worker well-being and improve the manifestation, in term of performance, of the system itself" (abridged from International Ergonomics Association 2016). Across this work, it is going to investigate the influence of constituent parts of the work context as shaping factors those manifest effects in performances. Contextualization in safety and reliability is here proposed as per multi-criteria judgments. It is going to construct a map of intra and inter level connections between human, working environment, type of task and t's coexistence

F. Ferraguti et al. (Eds.): HFR 2019, SPAR 12, pp. 154–185, 2020.
https://doi.org/10.1007/978-3-030-42026-0_12

factor of direct and indirect measurability. The map is elaborated pointing on concept map related with the physical, cognitive and visual fatigue reported in a stress main concept. The map is conceived as from a sectorial literature review and expert/managers guidelines. The proposal sheds light on the role of human factors on manufacturing and It contains robust links with easily enlargements in the industry4.0 behaviors (e.g., cobots systems (Fruggiero et al. 2019)). Moreover, this work maps the relations between what we decide and how system responds. The paper proposes a multi criteria decision making approach that set evidence of main factors affecting the dynamics of work-life and it reports human reliability analysis in manufacturing as per work stress. It was applied in a real manufacturing context, texting manual and tools assistant jobs, validated and analyzed with industrial managers and HF engineers.

1.1 A Literature Review Analysis

The role of workforce in supporting industrial objectives is recognized in many studies (Kolus et al. 2018). Humans interact with the environment conditioning the way by which the system is managed and reported in performances. Notwithstanding, workforce reacts to system under environmental, collaborative capacities and over productivity and equipment constraints (Fera and Macchiaroli 2010). Managers, since long time (Matteson and Ivancevich 1987), have been observing that the direct cost of unproductivity, large as they appear to be, are probably only a minor fraction of the true preventable costs of poor job design and poor operational strategy. In the most part of the applications, the human intervention is represented under a probabilistic approach. This depicts the likely possibility to report an accidental failure during the normal tasks due to the human intervention (Wittingham 2004). Humans report flexibility in systems but they generally affect optimization in performances (Sammarco et al. 2014). This is relevant whenever system has ambitious for leanness (Fruggiero et al. 2015). Human's actions are evidently a source of vulnerability and uncertainty for industrial systems. This got the scientific community to introduce, in the OM, the Human Reliability Assessment (HRA) which aims to deepen the role of the human factor in production. Over 90 human reliability tools have been developing since 1960 for the same thing: predict human behavior.

The HRA considers the probability that the human error occurs in a specific operation (it should change over time), when some causes, originated by the interaction of factors under fixed constraints, generate - or could potentially originate - a probability of failure in the routine of jobs (Di Pasquale et al. 2015). In the literature about HRA there is traditionally the contraposition of first and second generation approaches. The early techniques reported, in a whole, the classical risk analysis approaches: Probabilistic Safety Assessment/Probabilistic Risk Assessment (PSA/PRA) (Hirschberg 1990). The basis assumption is that because of natural deficiencies, humans logically fails to perform a task (Hollnagel 1996). This risk analysis, however, assesses humans can hardly be modelled as a mechanical/electrical component. He does not interact with other humans and machines and environment. Really, human affects and is affected over time by the performance requirements/conditions of production (Monsneron-Dupin et al. 1997; Moray 1998). Operator's reacts in time over standard factors (Apolloni et al. 2007; Fruggiero et al. 2016). The reliability is evaluated basing on Performance Shaping Factors (PSF)

(Furuhama et al. 1995; Di Giulio et al. 2000). PSFs do not exert their effect on performances but they act as stings of the different environmental states on humans. These are error producing conditions those require significant assessor judgment and mathematically get probability of error. So that, the Human Error Assessment and Reduction Technique (HEART) use PSFs under basic judgment reduction guidance (Williams 1988). Inside the techniques of the first generation the most popular are: *(i)* THERP (Kirwan 1996); *(ii)* SLIM-Maud (Kirwan 1992) and *(iii)* JHEDI (Kirwan 1996). Model of first generation concentrated towards quantification of effects instead of factors affecting human reliability. According to those, the human behavior is deduced from psychological and behavioral sciences (Holmberg et al. 1999). The first class models are limited at that very birth, *i.e.,* they originated by a cognitive mapping, they are mostly behavioral approaches and they were adjusted/adapted as per production engineering (Hollnagel 1996; Sträter et al. 2004). The need to respect a suitable cognitive model as per productivity assignment leads over the evolution of second-generation techniques. These get over to explain the relations and interactions between the humans and the technology/process. The second generation HRA aspires to conceptualize, using a taxonomy, human errors. It was studying in a deeper way the causes of the errors and their interactions, instead of their frequency in occurrence (Kim 2001). Marsden and Hollnagel (1996) investigated the human machine interaction using cognitive reliability and assuming that users cannot be predisposed (the term they used is "benign") in favour of task and technology. Second generation methods are (e.g., CREAM and ATHEANA) analysed and described in several publications (Hollnagel 1998; Cacciabue 2000; Kim 2001). They include human cognition as reaction and awareness in judgment. These suffer of a fairly small application range; they were born as per nuclear plant management, but it is worth to note that they are successfully applied into petrochemical, manufacturing and transport sector (Di Giulio et al. 2000). Their applicability gets value in those sectors where the cognition is limited, or can be, and benchmarked by manufacture and technical constraints. These restrict human unexpected reactions. Robustness remains the sore point. Vanderhaegen presented, under second generation models, a non-probabilistic approach to the ACIH (Analysis of Consequences of Human Unreliability). This aims at analyzing the consequences of human unreliability and the definition of adequate tools, for the prevention of errors by describing all possible scenarios (Vanderhaegen 2001), using a typical framework of the SCEnario Based Risk Assessment (SCEBRA). But, it requires accurate data collection (Gertman and Blackman 1994; Park 1987; Xiang et al. 2002) and it suffers in experience's formalization (Strater and Bub 1999). Examples of databases are the human event repository and analysis (HERA) (Hallbert and Gertmann 2004) and the human factors information system (HFIS) (Snodgrass and Szewczak 2002). To support reliability assessment, Structured Information Analysis (SIA) has been engaged (Jung et al. 2001). It tries to reduce the subjectivity of the analysts on the PSFs definition but it suffers in complexity of elaboration.

Most of the second generation model originated to classify somehow human errors (Basra and Kirwan 1998; Kirwan 1998; Richei et al. 2001). SOR (Stimulus, Organism, Response), borrowed from the psychological behavioral sciences, is introduced. SOR is assumed as the principle of human behavior: a stimulus acts on an organism to cause a response (Rau 1996). The failure to receive or the absence of a stimulus, incorrect processing of data acquired or an inadequate response can originate human error (Monsneron-Dupin et al. 1997). Slips, lapses and mistakes are all possible manifestations of human propensity towards tasks (Reason 1990).

Park, offers an extensive review of the possible causes of human error (Park 1987); in particular, he focused his analysis on two main errors causes, i.e., analyzing both execution errors and those resulting from wrong perception of an external stimulus. Moreover, he reported how experience positively interacts with the reduction of failures in execution (Nembhard and Osothsilp 2002). Second generation models inherit from first generation of HRA the classical division between errors of omission and commission (Dougherty 1997; Hollnagel 2000; Shorrock and Kirwan 2002; Sträter et al. 2004; Johnson 2014). It is worth to note that Hollnagel, developer of the technique CREAM, described quite extensively the differences and relationships between errors of omission and commission and he introduced categorization of error under contextual control Model (CoCoM) (Hollnagel 2000). That is:

- errors of omission: the lack of implementation of tasks required to achieve the result;
- errors of commission: the implementation of an operation not requested by the task and capable to generate an error.

Monsneron Dupin et al. 1997 defined DUA (Decision based on Unrequired Action). This model, based on cognitive learning errors, proposes to cauterize errors of commission differently from PSFs (Monsneron-Dupin et al. 1997). Perception, interpretation, planning and action are the components of the cognitive process that depends from working conditions as per the Standardised Plant Analysis Risk–Human reliability analysis (SPAR-H) (Blackman et al. 2008).

The novelty in the second generation of HRA is about the taxonomy of errors through the cognitive mapping. This has to be introduced to achieve the precise cognitive reaction that better describe human behavior at work (Dougherty 1998).

Despite a so general approach to the taxonomy definition, it is confirmed that human behavior introduces uncertainty. Pre-identification of relevant human factors is required over time as per the successful Nuclear Action Reliability Assessment (NARA) (Kirwan et al. 2004). But, Human generally reports failure in tasks that occur under random events (Rasmussen 1997). Failures may be controlled under cognitive perception (Village et al. 2013). Those who might insist that human performance is randomly originated from personal reactions, are prone to describe "likely" and "typical" reactions. These words remind the stochasticness of the process. Those who have interest in risk management are suggested to introduce probability in model rather than a precise predictability (Dougherty 1997). The application of the HRA models to the industrial environments (until discrete events stimulus) can be made if a mathematical characterization of the HRA is introduced. With regard to the assessment of human reliability, Park offers a wide overview of the probability distributions that best describe human behaviors in both discrete and continuous operations (Park 1987). These include the exponential and Weibull distributions. The common basis for all techniques is the evaluation of the HEP (Human Error Probability) that represents the index of human error and its modification is possible altering the influencing factors (PSFs).

Aim of this paper, as previously stated, is to present a probabilistic decision support model. It is capable to depict HF effects under PSFs on the industrial Operations Optimization Problem (OOP) assuming correlation between Work related stress and Human Error probability as per Adams 1980, Lazarus and Folkmann 1984, Wegner 1988, Reason 1990, Savery and Luks 2001, Colombini and Occhipinti 2004, Knisley 2005, Yeow et al. 2014. It is able to model, in the type of environment condition of work assisted and cooperation with robotics.

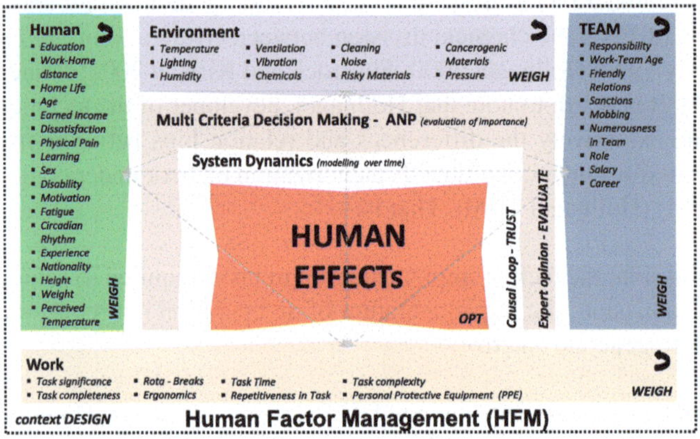

Fig. 1. Human Factor Management (HFM): a framework to save the effects of human at work.

2 The Methodological Approach

The Human behavioral science proposed a context in which humans at working begin to feel the pangs of action whenever the stimulus is over a threshold level. The threshold can be set according to personnel (Human), interaction (Team), requirement (Work) and context (Environment) in which is working. There is a kind of high-pass filter that reports reaction under stimulus over a particular amount. This level changes over time, mainly according to cumulative rules. It requires conceptualization and management strategies under pareto frontier analysis. The main Human Factor Management frame of this proposal starts from these assumptions (Fig. 1). Figure 2 reports the general steps required to model the humans interaction within system. It was centered in the evaluation of work related stress. It is based on a Deming approach (Scherkenbach 1986). We are reporting a main frame to evaluate the human reliability starting from the stress factor. First of all, the characterization of task, as part of the flow process, needs to be designed in terms of its ergonomic and operational requirements. In this part, line balancing principles could be considered as well as the ergonomic sets of the single task in the station. Tasks comes out from the context in the form of video, reports, designs, expert notes, analysts considerations etc.… .

Tasks are inserted in the jobs sequence (i.e., repetition and schedule priority) and come out with factors related to the context, workload, the activity, the process, the

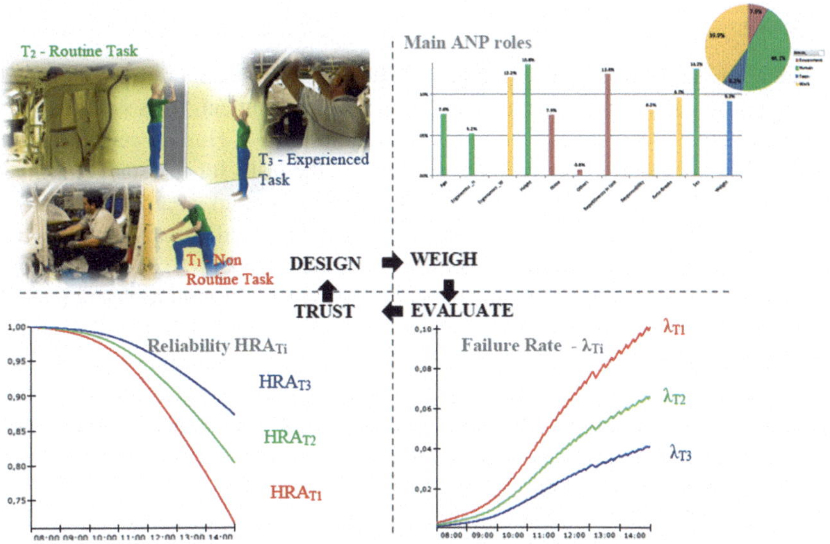

Fig. 2. Assessing the reliability of humans at work: a Deming approach to trusty human at work

personal cognitive intuition/propensity, the physical attitude, the personal ability, etc...... Factors originated stress in the form of physical, cognitive, visual fatigue. Clusterization in PSFs in required. They all contribute to the main STRESS outcome (capital letter as per main effect analysis of Fig. 3).

When all the tasks are designed on the line sequence, it is needed to decide which main factors influence stress. We started from the assumption that we cannot describe anything but we can eventually manage predominant manifestations while neglecting what is unimportant. We are conscious this approach will implicitly contain probability of error but It will gain in predictability. Main internal and externa l parameters that could alter (as from past experience and expert analysis) the failure state need investigation.

Sono factors can be reduced in parameters as measurable according with EMG, heart beats, galvanic sensors and images elaborations etc.... A decisional analysis is required in order to set the relative importance of factors influencing the work. A network map must be applied for prioritization. This requires consistency in elaboration, analytical comparison and hierarchical structure for decision. Qualitative values are included from experts, quantitative amounts are gotten from the environment and upon dynamics. These may eventually manifest failures in task over time. Then, main roles are highlighted. Their importance is weighted across quantitative analyses. Their dynamics is simulated over effects modelling. Consequently, the integral probability for the human reliability can be calculated. This is reporting the Human Reliability value used to assess and to trust the human actions in system.

In the following paragraphs, the method used to realize all the paths of this approach will be explained. A common route to world class performance is here applied (Deming 1986). Contextualization will be reported.

2.1 Design

The evaluation starts from the environment. It is required collection of data and to design about how the worker is collocated into the environment. This is mainly - in a design perspective - the work of an industrial ergonomist. A Human centric design is demanded. The first step of the method is to design the tasks for the stations, this step has to be realized having two main issues on the designers mind, *i.e.,* performances and the ergonomics. This phase is required in order to capture the dimensional benchmarks of the workstation, to report the anthropometry of humans at work, to collect test data, to set the robotic assistance/co-existence intensity. To do this, different tools are available. It can be included balancing between movement in sequences (mixed-model approach for line balancing problem and MALBP can be used) with ergonomics insights for the allocation of task in the flow line. 3D virtual tools can be used. External tools can be used for the ergonomics evaluation of the tasks, such as JACK®, Siemens PLM®, 3DSSPP, etc.... . A list of evaluation tools is available in the most recent books of Stanton et al. (2004), Wilson and Corlett (2005).

The results of the design phase are: the definition of work sequence, its repetition, the allocation of a particular humanoid at work, the characterization of task in term of load exposure, time flow, measurement of the geometry of task, the robotic interaction. Constants of the processes and measurable variables are, along this phase, quantifying. The interaction between dynamic events in system remains un-resolved.

2.2 Weigh

This part starts from the concept that two people can see the same thing and yet differ with each other (Covey 1989). Once that the process is designed and the line is balanced in terms of cycle time and ergonomics, it is possible to go forward the next step, *i.e.,* the weigh step (Neumann and Dul 2009). Each factor in system can alter the steady state reaction. The influence of a factor changes over time. Factors interacts each other. There is not the independence between element in system but a conditional influence in loop arrangements. Factors are "nodes" and their relationships (one node leads to or influences others) are considered "links". It is worth to say that factors can cumulatively report alteration in state. There is not the possibility to describe and predict all connections between elements in system but we can concentrate in what is supposed more important while neglecting something un-relevant. These are the principles of a percentile analysis. But, who has to decide? Decision is here reported in the form of relative weights. These are collecting from questionnaires, interviews, expert opinion, analyst, visual tools. They are elaborated on a hierarchical decision network. They are all included in the main model as per appendix amount.

This step is divided in two main blocks. The first one is the identification of the main parameters for the stress, and these can be selected through brain storming with all the direct workers involved in the line, as per the map reported in Fig. 3. Workers act including their own state (sex, age, anthropometry, knowledge, experience, predisposition, perception, civil state, etc.). The second step is to weight the main parameters previously selected through the brain-storming in Delphi approach. The analytical weighting can be done applying a well-known and reliable method such as the Analytic Network

Process (ANP) (Saaty 2008). This can guarantee consistency, hierarchy in decision, sensitiveness in factors. As following, ANP is briefly described for repeatability issue. The ANP will be applied to the net reported in Fig. 3. Here is briefly explained in its construction hypotheses. If we focus on human effect in terms of stress evaluation, to effectively assess the work-stress in the industrial operations, it is important to understand which are the main elements that can contributes to affect stress. There are some variables that comes from the safety guidance (reported also in Table 1), some from the operations and some from the ergonomics and work psychology. All these factors (constant, measurable, and no, variables) may be connected each other. Some factors are marked off (regulations may impose threshold- examples in Table 1) some others are free to set. The main elements to be considered for the stress phenomenon are referred to four main clusters, below reported:

1. the factors strictly related to the **HUMAN** behaviours,
2. the factors related to the work **ENVIRONMENT**,
3. the factors related to the work in **TEAM** and,
4. the factors directly connected to the tasks of the **WORK**.

Connection among elements of the same cluster is possible, connection between elements of different cluster is costumed. Intra level links are reported in dashed line, inter links across levels are designed in continuous line (see Fig. 3). All parameters highlighted have different impact on work-related stress. Indeed, there are variables those affect directly and proportionally the growth of the stress and others, even though they are not directly linked to stress, are affecting humans at work by influencing the variables linked directly to the work-stress parameter. The content of the map can be explored by looking out the most frequently linked concepts to human related factor in system. Map content identifies clusters in the analysis of proposal. Central concepts reflect perceived "influence". Connections can be analyzed qualitatively by following links originated by interview with managers, worker questionnaires, literature survey, HF engineers (totally 35 persons were involved). The possible connections are of two types: *(i)* intra-level and *(ii)* inter-level. The first group they share connections between parameters belonging to the same macro areas. These connections represent what generally is intended as direct influence among parameters and PSFs that can report in stress. Direct influences are generally studied in focus management initiatives and whenever boundary is missing. These are designed considering that stress principally evolves among channels. These have direct connectivity to effect. Notwithstanding, the direct stimulus works on bias amount. The stimulus privileges direct connections but it manages, in high pass approach, indirect affections. Direct and un-direct connections could have different weight on local effect even though they work with the same figure. Un-direct connection is the perceived memory/cognition and It is generally related to the individual. These elements are commonly ignored in optimization but are required whenever humans act in system. Each link in connection contains indicator that can boost or cut local effect. These arrows in map represent the direction of the relationship between the two linked parameters. It indicates which of the values is the one that affects or, conversely, is affected by the other. Closed Loop Map of Fig. 3 is designed to allow, each of readers, to create their own stress map, that has to be arranged for each specific application field.

Clusterisation starts with relative judgments (i.e., "weight" phase) as to recognize the main factors influencing the stress in the work, through a brain-storming meeting. They, starting from focus learning groups, need to correlate the designed tasks and operations of work environment. It is required to identify the main elements on which experts are called to express their own judgment for the weight calculation through the ANP application (because judgments need consistency, clusterization, moving in average, hierarchy, counting, remarkableness). Let us, now, briefly explain what ANP approach is.

Table 1. Main safety factors influencing the work-stress with limit values and good practices of exposure

Safety factors	Norm
Video-terminals	Break every 15 min each 120 min working. Screen minimum technical requirements, the work desk characteristics, equipment, lighting and micro-climate
Noise	Exposure value limit 87 dB(A), limit instantaneous peak value exposure 200 Pa (referred to 140 dB(C) with 20 μPa)
Temperature	The temperature in working areas must be adequate for human beings during working hours, in view of the working methods used and the physical demands placed on the workers. One must also taking into account humidity and ventilation. Temperature limit changes in relation to perception
Lighting	Workplaces must have sufficient natural light. All workplaces must be equipped with artificial lighting adequate to protect the safety, health and welfare of workers
Air change	Workers must be guaranteed sufficient and healthy amount of air through natural ventilation or mechanical ventilation
Manual loading	Load characteristics, the physical effort required, the space available and the individual risk factors
Hand-arm vibrations	The daily exposure limit, normalized to a period of 8 h, is 5 m/s^2; of short periods is of 20 m/s^2
Body vibrations	The daily exposure limit, standardized to an 8-h period, is 1.0 m/s^2; over short periods is 1.5 m/s^2
Radiations	For wavelengths between 80 and 300 [nm] the value of daily exposure is 30 [J/m]. For wavelengths from 300 to 400 [nm] the value of daily exposure is 104 [J/m]. For wavelengths between 400 and 3800 [nm] the value of daily exposure is calculated from the following relation 20,000 * $t^{0.25}$ [J/m]
Chemical compounds exposure	The exposure limits are reported for each product used on the safety sheets of the chemical products accordingly to the European norms

Based on the contribution of different experts, it aims at the creation of a unique priority index for each possible decision, that summaries all expert's judgments, minimizing their inconsistency. Generally, the procedure, given an objective and a set of possible choices and/or decisions to achieve the goal, calls the experts to express a relative judgment of relevance of each choice, when compared to all the others. So, given a set of possible decisions, $D = [D_1, D_2, \ldots, D_n]$, the expert has to indicate a relevance judgment of each decision compared with all the others, examined one by one. Each expert gives a relevance judgment, that could be named j_{kil}, where k and i are the counter of all the decisions belonging to the set D and l is the counter of the *l-th* expert. The geometrical mean used for the collection of different judgments is fundamental, because it has been demonstrated (Saaty 2000) that using the geometrical mean minimizes the inconsistency of the comparison matrix C (see forward), for the reason that the unanimity and homogeneity properties are respected. All judgments for each couple of decisions (D_k, D_j), will be synthesized using a geometrical mean through the (1). The unanimity property states that, when all the experts give the same judgment x, the resulting overall

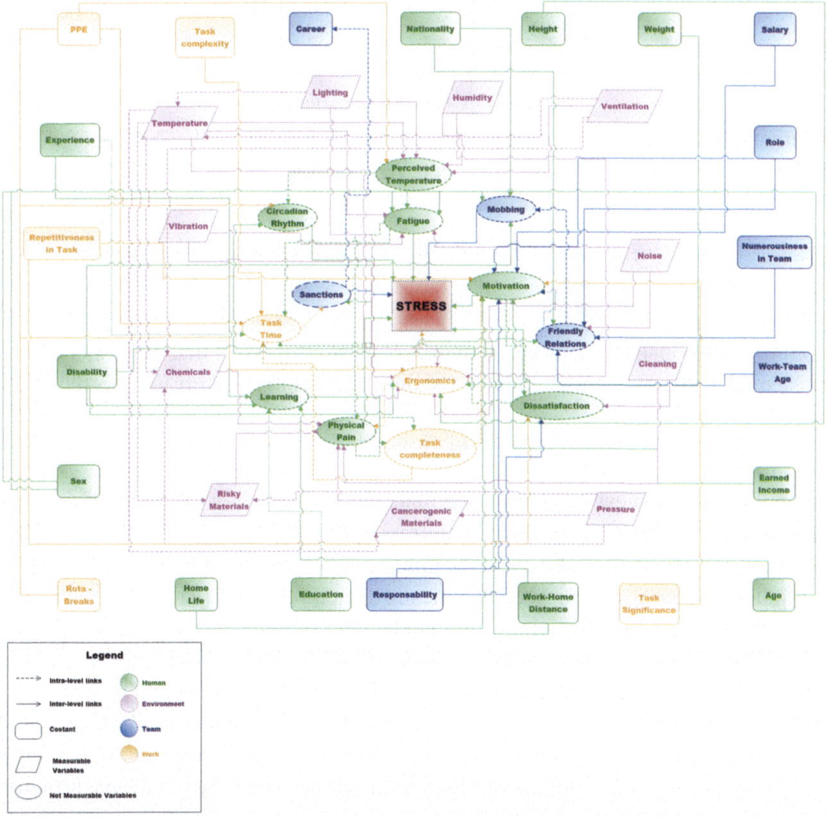

Fig. 3. Network for stress analysis in tasks - (the strategic goal is the red block - the term STRESS is here defined as per the human main effect to reliability assessment)

judgment should be x.

$$j_{ki} = \sqrt[n]{j_{ki1} \cdot j_{ki2} \cdot \ldots \cdot j_{kin}} \tag{1}$$

The homogeneity property states that when individuals give a judgment u times larger than another, the resulting overall judgment should be u times larger too. A mathematical formulation of the unanimity and homogeneity properties is reported in (2) and (3).

$$f(x, x, \ldots, x) = x \quad \forall \quad x \in X \tag{2}$$

$$\begin{aligned} & f(ux_1, ux_2, \ldots, ux_n) = uf(x_1, x_2, \ldots, x_n) \\ & \forall x, ux \in X, u \in R \end{aligned} \tag{3}$$

Once the resulting overall judgments are computed, through Eq. (1), they are inserted into a square matrix (nxn), named comparison matrix, C.

$$C = \begin{array}{c} \\ D_1 \\ D_2 \\ .. \\ .. \\ D_n \end{array} \begin{array}{c} \begin{array}{ccccc} D_1 & D_2 & .. & .. & D_n \end{array} \\ \begin{pmatrix} j_{11} & j_{12} & .. & .. & j_{1n} \\ j_{21} & j_{22} & .. & .. & j_{2n} \\ .. & .. & .. & .. & .. \\ .. & .. & .. & .. & .. \\ j_{n1} & j_{n2} & .. & .. & j_{nn} \end{pmatrix} \end{array}$$

A fundamental condition to be respected for the applicability of the ANP methodology is that the comparison matrix C is consistent. We say that a matrix $A = (a_{ij})$ is consistent if the transitivity and symmetric properties are satisfied, as expressed in (4) and (5).

$$a_{ij} \cdot a_{jk} = a_{ik} \quad where \quad i, j, k = 1, 2, \ldots, n \tag{4}$$

$$a_{ij} \cdot a_{ji} = 1 \quad where \quad i, j = 1, 2, \ldots, n \tag{5}$$

To maximize the consistency index of matrix C, besides the fact that elements with $k = i$ always equal 1, the elements with $k \neq i$ should respect the condition reported in (6).

$$j_{ik} = \frac{1}{j_{ki}} \tag{6}$$

In real application, however, it's possible that relation (6) is not satisfied. This implies that an inconsistency may occur. For the correct applicability of the ANP methodology, it's important that the inconsistency of the comparison matrix C is less than 10% respect random judgments, *i.e.,* the number of times in which relations (4) are not verified has to be less than 10% respect random judgments. The ranking of the possible decisions D_i, as stemming from the judgments of experts and from the field, can be computed from the entries j_{ikl} of the comparison matrix C. In particular it is needed to calculate the maximum eigenvalue λ and so the relative eigenvector of all the matrix $v\lambda$ (Saaty and Hu 1998). After this calculation, normalizing the eigenvector (as shown in (7)), obtaining a percentage judgment, it is calculated a final value of priority (Fig. 4) for each factor of decision, i.e. for each row of the matrix C.

$$p_i = v\lambda / \sum_i v\lambda_i \tag{7}$$

2.3 Evaluate

Direct connection (bold lines) and un-direct connection (grey lines) are reported as per the mutual influence in Fig. 5. A Causal Loop Diagram is constructed in order to test the failure over time. The is based on a holistic vision that reports about causes of human. The connection between factors (i.e., direct and un-direct links) is created over literature and expert revision. Weights between and across connection are reported - as per Fig. 4 - after the ANP analysis with the intent to report human involvement in error (Fera et al. 2017). The CLD is synthetically sketched in Fig. 5. We are interested here in the context of results of the diagram: to relate states to failure rate. It results in the timely failure rate during task as per context and under different working requirements. System evolves - in causal loop - under links between states, relation among factors, and connections across "constant" parameters. Links have arrowheads that show the causative direction.

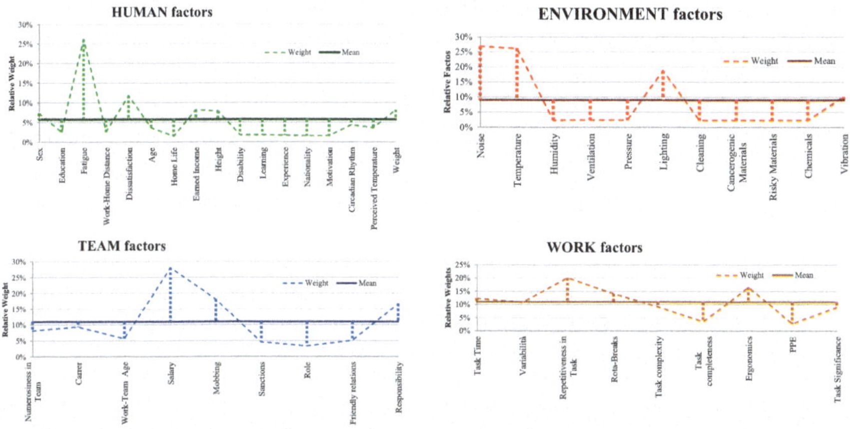

Fig. 4. The weighting results for each single concept into the system as from 2 academicians, 1 doctor, 2 industrial engineers, 30 workers judgments in the cases study under investigation. Consistency of results as after ANP elaboration. Case studies form a manufacturing elementary technology unit.

It dynamically simulates the failure rate and cumulative probability to failure under those assumptions (i.e., DoE plan). Dynamics of system is reported. Measuring the work related stress in a particular context and under main stimulus from the Human, Environmental, Team and Work factors is the aim as this part. The map of Fig. 5 requires interpretation in terms of which element contribute to ..., what is the effects of that element. In particular, it s possible to consider three kinds of blocks in the work-stress map:

1. Constant parameters (rhombus shape): these are constant factors - they depend from the context and are set in DoE plan - (*e.g.*, sex of the operator, operator age, number of people in the team, etc.). The decision-maker can fix constant in order to simulate reactions.

2. Connections: these are direct (black) and indirect (grey) relationships, respectively, among and between paths from main states.
3. Measurable Variables (red circles): each state depends from the environment and the context (which can be described from constant parameters) but it reacts to the manifestation of experimental variables (measureable ones). These variables act, more or less, like factors and they require continuous monitoring and controlling (e.g., EMG, thermal images, phomemeter reports, CO_2 amount, electrical reactions using galvanic sensors). They can change over time and across context (e.g., internal "temperature", "humidity", presence of "carcinogens", "lightness level", "Heart rate"etc.)
4. Variables not measurable (green circles): they are not an evident manifestation and it is difficult to measure, directly, their amount over time. They cannot be quantified without a design perspective. Such variables mainly depend on direct and un-direct connections across and between factors and parameters and constants that describe the system. These are main factors. They could be controlled through continuous survey (e.g., task *Complexity*, *Training* effects on *Workload* etc.…) and global judgment. They are generally are defined across tables of judgements;
5. States, (yellow circles): they are the system states - effects. These are the aggregate manifestation of cumulative factors those are affecting the work as per human interference. These states have direct and un-direct connection between and across and among clusters. They are mainly consequences of all other nodes previously described;
6. Failure rate (the hatched circle): it is the final element to be evaluated/calculated, i.e., the failure rate for the humans' interference in tasks. This changes over time and It could act form an initial condition according to the accumulate amount of stress to work process contribute to generate.

All parameters have positive (P) or negative (N) or both (P/N) influence, in direct and un-direct connection, with all others node in CLD. Whatever the directional connection is P, the source's growth will improve the output in term of performances. On the contrary, a N connection implies waste in state, i.e., it will contribute to improve the failure rate. Notwithstanding, there are some factors that could contribute in reduction to failure but they can invert effects as times change or boundaries is muting (i.e., repetitiveness in tasks - learning and forgetting rate are there involved - as well as responsibility - it could acts as stimulus or depressive force as per mental cognitive synapses). A system dynamic modelling is used for simulation (Sterman 2000). Namely *In Factors* are under the direct control of the particular workforce amount in process, unlike, *Out Factors* are assumed as fixed by the process in the testing experiments. Authors are assuming, in the CLD diagram, that slept hours could influence the circadian rhythm and those report personal endogenous factors (*In Factors* as per CLD of Fig. 5). From the analysis of literature review, there is the absent of direct correlation between environmental variables and circadian rhythm. Experimental studies were made but generalization in the field is inopportune (Aschoff 1965; Ivanov et al. 2007; Lerman et al. 2012;).

Rota and *Break* are included under the class of Not Measurable Variables as they depend from the strategy of management - so managers can fix amount - but, in the context of map of Fig. 5, they report a variable effect that can be qualitative related to

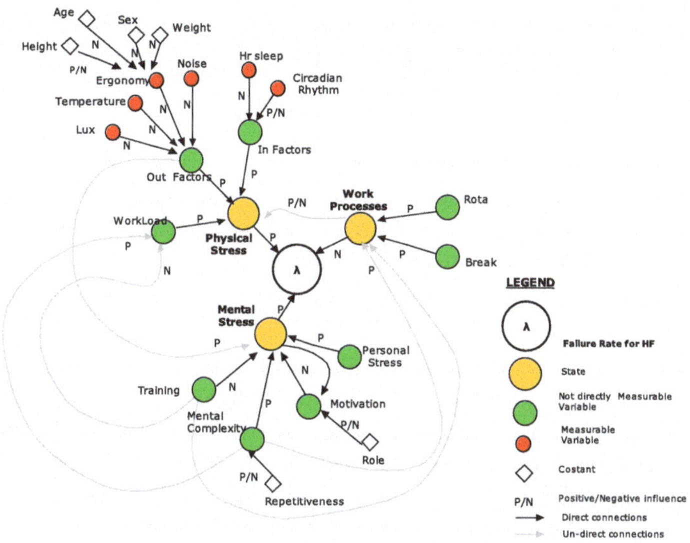

Fig. 5. Causal Loop diagram for failure manifestation in work related tasks – main relationships. Elements of maps are quoted in text under own typing (Bold Style for State, Italic type for Not directly Measurable Variable; in inverted commas for Measurable Variable, underline for Constant).

Work Process state event though they influence the **Physical Stress** state (Di Pasquale et al. 2017). According to the production strategies and basing on the schedule pressure, it is possible to include a state that report work-process effect in system. It report the amount of the schedule pressure the manager is setting on the working day. It is not measurable, although checkable, how much *rota* and *break* contribute to failure (Brown et al. 2010), how much robotics can alter the cognitive fatigue. They change over process and they manifest influence in physical fatigue for example.

Work Process interferes with the physical stress because it mainly gets the amount of physical fatigue the worker reports. It is revealed over workload requirements. **Physical Stress** is mainly directly dependent by the type of task the operator is performing (i.e., **WorkLoad**), the attitude of worker to be focus on task (i.e., *In Factor*), the type of workstation and system/process conditions under d assignments (i.e., *Out Factor).* In order to model the **Physical Stress** state in system dynamics, it has been decided to assumed relations among variable as from the main bibliography in the field. The model respects the current vision: awkward posture, high force, high repetition, long durations may lead to work related musculoskeletal disorders and consequently they represent risk factors (Gallagher and Heberger 2013). The concept of **Physical Stress** is widely discuss in literature Whittingham 2004. It starts from the psychological and physiological and ergonomics domain. A worker that is performing a task, maintains a Maximum Endurance Time limit (MET) (El Ahrache et al. 2006). The **Physical Stress** has the main effect in fatigued worker. The amount of MET - the point at which individuals can no longer exert the required force level (Rohmert 1973; Niebel and Freivalds 2003) is under the effect of exertion the particular worker is performing. MET is evaluated as fraction of the individual's maximum capability (f) under continuous static loading conditions (Van Dieen and Oude Vrielink 1994).

$$MET = Be^{-\beta f} \tag{8}$$

where B and β are static model-specific parameters.

The fatigue index has been conceived as the result of the time (MET) x load impulse force (Maximum Load Capacity, i.e., MLC). The maximum Fatigue index (L_{max}) for the task *i* can be consequently statically determined

$$L^i_{max} = MLC \times f_i \times MET_i \tag{9}$$

Thus, according to a particular task we are able to define MET_{ij} and f_j and MLC_j, where *j* is the type of anatomic joint we are taken in consideration. This study assumes the dependence from joints in a constant mode so, an average across joints value as been assumed in calculation. It is evident here the dependence of physical stress from constant and measurable and environmental variables into the system. The tasks to be performed and the total limit of maximum fatigue depend from the bottleneck.

$$L_{max} = \min\{L^i_{max} | i = 1, \ldots, n\} \tag{10}$$

Whenever in a system is going to perform different tasks at different load conditions and under different time dependent state, the amount of physical stress the worker is storing up can be defined as L_T after time *T*

$$L_T = \sum f_i t_i - a_i b_i \tag{11}$$

Where t_i is the length (of time) of task *i*, b_i is the break time following task *i* (in cycle *j*) and a_i is the allowance rate defined in

$$a_i = \frac{L_{max}}{RA_i \times MET_i} \tag{12}$$

With RA_i (Rest Allowance) required for task *i* as fraction of MET_i (assumed as $3 \times MET^{-0,41479}$ - (Konz and Jonhson 2004; Iannone et al. 2004)).

It has been assumed worker completely fatigued after performing of task of 30% of their maximum Load capacity after 75 no-stop time units and a Recovery Rate equal to 50%. These are constant factors in model. In the Eqs. (13 and 14), task time t_i is required. Time depends from the workload. It is un-directly connected with training and mental complexity. The more are the times (*k* - repetitions) the worker is performing the task *i*, the lower is cycle time (Wright 1936)

$$t_{ik} = T_{i1} k^{-l} \tag{13}$$

Where t_{ik} is the time required for the *i* task after *k* repetitions and *l* is the learning rate worker report. This means it is needed a monitoring and controlling step where repetitiveness can be assumed as constants (assigned at the work day). Starting from learning, after $m \geq k$ repetitions workers reacts with forgetting rate *f*.

$$t_{im} = T_1 m^f \tag{14}$$

Forgetting is a worker related effect. It could depend from mental attitude and task complexity. It is will act improving fatigue and, consequently, boosting failure rate in its **Physical Stress** state.

This system dynamics block is reported in Fig. 9, data of tasks as per Table 4. The amount of MET_i and MLC depends from the "ergonomy" in design. There is generally not one index that can completely describe "ergonomy". Authors decided to use: *(i)* NIOSH - National Index Of Occupational Safety and Health - to measure the RWL (i.e., Recommended Weight Limit - RWL) in LI (i.e., Lifting Index) for that particular task under those humans at work (constant factors in their anthropometry) (Waters et al. 1993); (ii) Snook & Ciriello (Snook and Ciriello 1991) to analyze stress and strain limits with definition of LBD (i.e., Low Back Disorders) as PSF risky factor (assumed based on static report) in repetition; *(iii)* RULA - Rapid Upper Limb Assessment - in posture analysis (McAtamney and Corlett 1987; McAtamney and Corlett 2004). Those require *Out Factor* modelling and composition based on expert judgment and prioritization (weights). They are Not directly Measurable Variables in the sense they need a design phase in order to be quantify. Borg scale is used to measure/weight the perception of exertion human manifest during task (Borg 1990). Interview are required. Medical consultation is demanded. Thus, MSD (i.e., Muscle Skeletal Disorder) index is finally reported (Nordin and Andersson 1997).

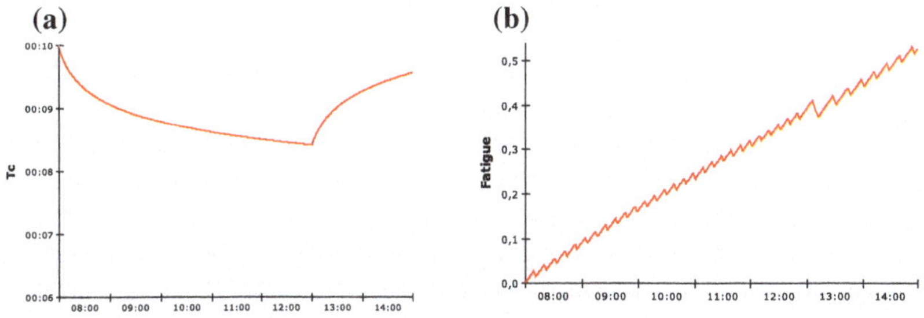

Fig. 6. Reporting the dynamics of cycle time as per learning and forgetting effect (Fig. 6a) and the fatigue cumulative load (Fig. 6b) over turn in shift.

Author decided to aggregate these three ergonomic indexes (Fig. 9 as per dynamics modeling block) because they - all merged - are able to report about load posture and repetition and dynamics as per overall human reliability assessment. These are quantified in the "ergonomy" measurable variable of Fig. 5. For the sake of completeness, and to the modelling aim, a synthetic description is here included.

The first index NIOSH was measured according to the Eq. 15 based on the simulation reports and measurable variables related to the context.

$$RWL = LC \cdot HM \cdot VM \cdot DM \cdot AM \cdot FM \cdot CM \quad (15)$$

where: $LC = 23$ kg - i.e. Load Constant; $\underline{HM} = 25/H$ - i.e. Horizontal Multiplier with H horizontal location; $\underline{VM} = 1 - (0.003|V - 75|)$ - i.e. Vertical Multiplier with V vertical location; $\underline{DM} = 0.82+(4.5/D)$ - i.e. Distance Multiplier with D vertical travel distance;

$\underline{AM} = 1 - (0.0032A)$ - i.e. Asymmetric Multiplier with A asymmetry angle; FM - i.e. Frequency Multiplier depending upon the average lifts/min F; CM - i.e. Coupling Multiplier on good (1.00), fair (0.95), poor (0.90) scale.

The Snook and Ciriello (S&C) indexes assumed as from database of tables and depends from the envornment. Limit values reported in Snook and Ciriello requires experimental tables and studies on the musculoskeletal load, pain, endurance and fatigue associated to pushing and pulling tasks. It requires classification based on operator gender, distance, height handles, frequency. Data are reported under 90% of adult and healthy working population. It such a study, under that working condition, S&C is a measureable variable gotten from costants.

The RULA model assigns a score to the postures of several body parts; afterwards the total score is calculated. This score, which ranges from 1 (good posture) to 7 (very critical posture) returns the urgency about the need to change how a person is working as a function of the degree of injury risk. Point are added if the posture is held for longer than 1 min or repeated more than 4 times per minute. The rula pace is associated with a particular level of intervention priority. RULA is the report of a virtual simulation study. The index recognizes four action urgency levels subdivided as follows:

- level 1 (score of 1–2): the person is working with a good posture with no risk of injury from work posture;
- level 2 (score of 3–4): the person is working with a posture that could present some risks of injury from work posture, and this score, most likely, is the result of one part of the body being in a deviated and awkward position, so this should be investigated and corrected;
- level 3 (score of 5–6): the person is working in a poor posture with a risk of injury from work posture. The situation needs to be investigated and changed in the near future to prevent an injury;
- level 4 (score of 7): the person is working with a very bad posture with an immediate risk of injury from their work posture. The situation needs to be investigated and changed immediately to prevent an injury.

Environmental factors, besides from ergonomics in workstation, act on system in the form of context based variables. There is generally, in literature, the lack of an analytic approach to describe the **Physical Stress** related to the out bounds (Aschoff 1965). Causal diagrams may be modelled according to experimental reports (Czeisler and Gooley 2007; Ivanov et al. 2007; Lerman et al. 2012). Based on legislation and experimental data it is possible to assume a class based scale of influence. In the CLD diagram "Lux" and "Temperature" and "Noise" act as *Out Factors* ("ergonomy" is inside but was differently modelled, as previously discussed) weighted on Borg scale based on range modularity effects. They act as physical hazards as per Occupational Safety and Health Administration (i.e., OSHA) topics.

Out Factors interacts with *In Factors* to originated human **Physical Stress**. *Out Factors* state contributes to alter **Mental Stress** *i.e.,* cognitive effort.

The "Circadian Rhythm" is influenced by phase and amplitude of "Lux" and "Temperature" and "Noise". They alter the standard individual's "Circadian Rhythm" in terms of its phase and amplitude. *Out Factors* and *In Factors* in this form originate **Physical**

Stress amount. Deviation from standard directly influences the **Physical Stress** and obviously the mental cognitive/concentration capability. The more is the deviation the higher is the effect. A faster phase gets a very quick changing in moods. The "Circadian Rhythm" changes with "Hr sleep", that is the amount of slept hours, for its direct manifestation in *In Factors*. A significant alternation, in terms of amplitude, was registered in people with lower than 6 solid hours in slept per day (individually change the minimum requirement of hours/day). Sleep disruption may lead to fatigue (Guyton 1992). To sleepiness follows inattentiveness. Moreover, between the tenth and twenty-sixth hours of wakefulness, mean relative performance on the tracking task decreased by 0.74% per hour. In order to understand the role of such a variable, the performance decrement for each hour of wakefulness between 10 and 26 h was equivalent to the performance decrement observed with a 0.004% rise in blood alcohol concentration. (Dawson and Reid 1997). An estimation of failure rate in assembly task is reported in the work of Lombardi et al. 2010. The CLD concept is modelled in Fig. 10 considering data as per Table 2.

Table 2. Failure rate as related to sleepiness.

Estimation of failures rate/100 workers across weekly (31–40) work hours							
Slept hours	<5	5–5.9	6–6.9	7–7.9	8–8.9	9–9.9	>10
Failure rate	7.89	5.21	3.62	2.27	2.50	2.22	4.72

All the factors, as above described, are under the notation of **Physical Stress** and they require individual characterization and context based measurement. Increase in work-related injury risk was implemented with decreasing - usual daily self-reported - slept hours and increasing weekly work hours, independent from industry, and dependent from role, type of pay and education which are modelled in *Motivation* state of Fig. 12, sex, age, and anthropometric factors. Moreover, the **Physical Stress** was finally related to the type of task and amount of work the task requires. *Workload* was measured based on the schedule pressure that is an auxiliary variables at time *t*. It is analytically determined based on the relation between what it is planned to be made and what was effectively produced. Work completed at time *t* is the level of production. Under the measure of a *Workload* amount there is obviously a characterization of time. It is clear the dependences of *Workload* variable to the repetitiveness as per Wright (1936) characterization.

$$schedule\ pressure(t) = \frac{schedule\ work\ completed(t)}{Work\ Completed(t)} \tag{16}$$

It has been assuming a direct relation between probability of injuries and requirement in production rate as per schedule pressure. A final **Physical Stress** state is reported in order to report about effect related to the above measured conditions.

Mental Stress is influenced by *Out Factors* and it is obviously related to personal attitude and visual fatigue. Stress in terms of cognitive capability may be related to the repetitiveness in task. The more the repetitiveness increases the higher alienation will be reported (Ramaswami et al. 1993). They indirectly influence the *workload* variable in system. *Personal Stress* may be an inherited condition in the sense it is influenced by out of production factors. It links personal life and working climate. It was considered as Not-Measurable Variable - it is not possible to directly quantify its role in failure - and reported from self-consciousness (Borg scale and ANP procedure are used). If work is performing in group, the CLD diagram is completed with the information about social interactions. It has been assumed that social relation - between people in group, i.e., elementary technological units - improves the probability to get correct decision (Steiner 1972). Based on the single contribution to decision (i.e., *A* - this value depends from type of task and type of work is performing), the probability to get a correct decision in a group of *n*, identically contributing, elements could be $P_0 = A^n$ and consequently the probability to failures is $P_F = 1 - P_0$. The group collective intelligence is strongly correlated with the social sensitivity in group (Hackman and Vimar 1972).

Notwithstanding, the higher is the group number the more complex are relations between elements. As it is not possible to include the quality of relations in group as well as the effect of such to failure - we decide to include weights among variables to monitor such uncountable factor – it has been assumed that there is a limit - based on experience and experimental studies (Riggs and Knight 1994) – at numerousness in group when a failure will manifest in 95 percentile (Jorgensen 1973). This limit in numerousness is used to normalize the effects of iterations. The model was experimental tested and verified in the context of analysis.

The n.ro of iterations is based on the combination of elements in group according to the followed relations. Collaborative capacity of teams is characterized by at least two components: *(i)* an attribute-based perspective that focuses on capabilities of single actors in organizations and their ability to bring intelligence into the decisional process, and *(ii)* a perspective on group dynamics which describes how interactions in team grows up with numerousness. According with the normalized amount of iteration and based on the single contribution to the decision process it is possible to assume an optimal definition of people in group. It is decided to use norm as per 11 persons in team and according to the experimental test of Lawler et al. 2011.

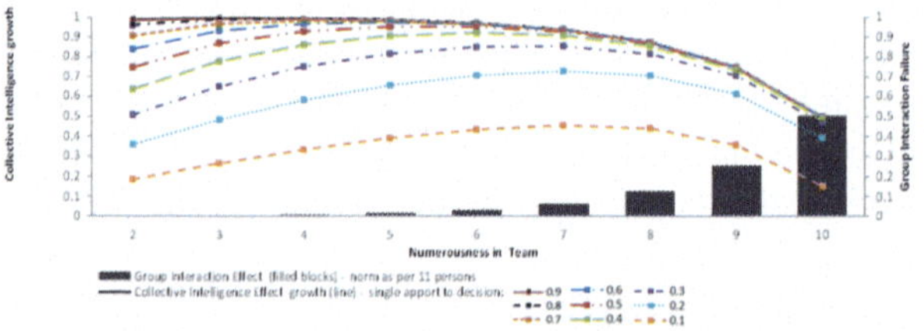

Fig. 7. The causality effect of team numerousness under different collective intelligent effects

$C_{(n,x)}$ defines *x*-combination of a set of *n* persons.

Where $x = 0$ means that group will have one organic external interference. The deviation between this number will proportionally act as failure rate.

It is evident (see Fig. 7) that if the single contribute to decision is over the 0.7% it is preferable a condition were the numerousness in team is under the 4 units in order to have no growth of group interaction failure probability. Lower than 0.6 single contribution to decision requires a minimum of 7 units in team in order to get a collective intelligence but this is moving up the group interaction probability of failure which in turn is related with the possible combination of relations among people in team (slips, lapses and mistakes, as per Reason 1990, frequently occur). *Training* in process will influence the time required in tasks. *Mental Complexity* is connected with repetitiveness. It has been assuming that repetitiveness is a constant, in the sense it can be included in the strategic parameter and demand requirements, while *Mental Complexity* is influenced by its amount.

Motivation is modelled on uncountable amount (we weighted on reports we collected and on queries we received in that particular industrial context). A relation between responsibilities and wage as well as role in system is considered as per CLD model.

2.4 Trust

After that the failure rate is calculated, it will be possible to elaborate the cumulative probability of errors for humans. The reliability function is reported in (8) and it is a negative exponential depending by the failure rate function defined in the time span of the shift (Di Pasquale et al. 2015). The function is the weighted mean (may be related to single task occurrence) along tasks for the systems subject to failures.

$$R = e^{-\int_0^t \lambda(s) \cdot ds} \tag{17}$$

3 Managing the Behavior of Human at Work: A Safe Based Reliability Approach

3.1 Inherit the Design Context

The purpose of the following illustrative example is to demonstrate the selection and use of multi-criteria decision with causality relations for the assessment of human reliability under stressed conditions (properly the approach of Fig. 1 as reported, per cases study, in 2). The experimental campaign was conducted in an automotive company and in particular on three human tasks of an assembly line division.

The team for the implementation of the method proposed in this paper was composed by 30 workers (all belonging to the assembly line analyzed), 2 health and safety engineers of the same company, 2 experts with experience in academics and 1 occupational physician internal industry. These are all signed as experts in the following sections.

Data were submitted in the form of questionnaire and collected in database (see appendix 1). Tasks (we graphically presented jobs in Fig. 2) have been classified in order to their repetitiveness as Non Routine = T_1 (*i.e.*, red line) and Routine = T_2 (*i.e.*, green line) and Experienced = T_3 (*i.e.*, blue line). Main roles in ANP (covering the 98.6%) have been reported in the bar chart (as per the ANP of Fig. 4) of Fig. 8. Expert affirmed that the **HUMAN** criteria is responsible for the 44.1% of stress in map. The particular **WORK** under investigation (we are talking about a production rate of 55 unit/h in mean among tasks) influences the perception of stress that is about 39.9%. **ENVIRONMENT** (in the investigated tasks, as per the manufacturing division of an automotive firm) and the **TEAM** clusters are responsible in stress of nearly the 8%. They do not manifest singularities in management. It is important to note that between all the parameters, the work team indicated the following as relevant for the stress in their work environment and reality. The factors are 9 and chosen between the ones present in the net of the Fig. 2 and they are: (1) Ergonomics; (2) Worker Height; (3) Sex; (4) Task repetitiveness; (5) Pauses in the shift; (6) Worker weight; (7) Responsibility and role; (8) Noise; (9) Age. These were modelled in the system dynamics map.

The three tasks are: (T_1) mounting soundproofing interior vehicle; (T_{12}) mounting wiring rear vehicle; (T_3) mounting seal front door vehicle.

Table 3. Prioritization for stress on main factors [%].

Type	%	Type	%
Ergonomics	17.40%	Weight	9.20%
Height	13.20%	Responsibility	8.20%
Sex	13.20%	Age	7.60%
Repetitiveness in task	12.60%	Noise	7.50%
Rota-breaks	9.70%	Others	1.4%

The tasks T_1 and T_2 and T_3 have been designed (in order to detect measurable variables) in a simulated environment and, after a weighted supermatrix elaboration, its performances in terms of stress have been evaluated. A 95 percentile has been used in design the ergo manikin. The condition of stress has been related to a failure rate and simulated - according to a system dynamic approach during the shift. System Dynamics reported the outcome of non-measurable effects.

For these three experiments, it was investigated the prioritization index for each factor that influence the stress in the opinion of the experts and of the workers directly involved

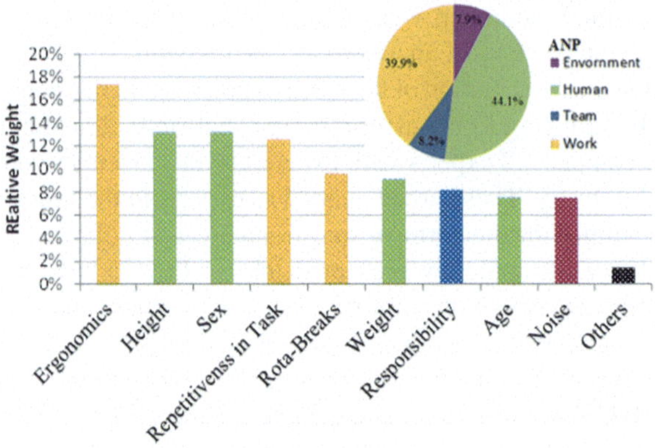

Fig. 8. The stress main factors prioritization

in the three tasks. The prioritization list is the one below (Table 3 and Fig. 8). The factors prioritization is very important for the stress evaluation and, so, for understanding of the human factor in the three tasks analyzed in the paper. What it is possible to understand from the values in Table 3 is that the main factors covering more than the 60% of the total importance for the stress evaluation is covered by: (i) ergonomics in the task execution, (ii) height of the worker, (iii) sex of the worker and (iv) repetitiveness of the task during the work-shift. The other factors have less importance in the evaluation of stress. The prioritization indexes were calculated applying the ANP technique, those values were applied to the dynamics closed loop diagram of Fig. 3. Thus, it was possible to calculate the failure rate of the humans involved in the three tasks, that are reported in the Figs. 5, 6 and 7. The three tasks simulated are characterized by the tasks values, reported in Tables 4. The failure rate is then been connected with a reliability measure - $HRA = e^{-\int_0^T \lambda(t)dt}$. The measure of reliability in task as from the stress map has been reported (Table 5) and considered for optimization in design. As it is evident, if the task has been performed by experienced worker, there is less probability to incur in safety issues. Probability increases whenever long repetition are imposes. When performing non routine tasks, strategies of work enlargement are preferable if we want to improve reliability of nearly 20%. The routine tasks could eventually incur in repetitiveness effects. In the proposed issue, "Ergonomics" (17.4%), worker Height (13.2%) and Sex (13.2%) and Weight (9.2%), Repetitiveness in task (12.6%) and *Break* in shift rota (9.7%) cover 75 percentile in influence. the main factor that can be arranged in order to manage the stress and consequently the reliability in system.

In Table 5 it is reported the results of the CLD for different hours of work and for each tasks before introduced. In the Figs. 9 and 10 it is reported respectively the mean across tasks failure rate and HRA curve. The shape is very well approximated with a cubic polynomial function, that has a square fitting index equal to almost 100%. The failure rate was, ended, calculated for different tasks belonging to the same assembly line, it was possible to calculate the final curve for the Human Reliability Analysis belonging to the specific assembly line and to the specific three tasks analyzed in these paper to demonstrate the effectiveness of the method here introduced. In the Fig. 10 it is reported the curve for the HRA as calculated through the formula (17). Thus, from HRA curve it is possible to understand that the human reliability, after the 6 h of work, falls dramatically down, lower 85%, that can have not just influence on the operations and product quality, but mainly on the safety issues related to the operations and tasks execution.

Table 4. Task characteristic in process under investigation

	Task n. 1	Task n. 2	Task n. 3
Task repetition per hour	81	36	49
Cycle time with pause [min]	0.74	1.67	1.22
Real cycle time without pauses [min]	0.65	1.50	1.00
Pause time span per cycle [min]	0.09	0.17	0.22

Table 5. The failure rates for each task executed in the work shift time span

Time	λ_1 (t) * Δt * 10^{-3}	λ_2 (t) * Δt * 10^{-3}	λ_3 (t) * Δt * 10^{-3}	$\lambda_{W\_mean}(t)$ * Δt * 10^{-3}
8:00	0.00	0.00	0.00	0.00
8:45	4.00	6.50	4.90	5.15
9:35	7.40	9.00	7.90	8.11
9:50	11.00	16.00	15.00	13.99
10:10	18.00	23.00	19.00	20.05
10:20	24.00	25.00	24.10	24.38
10:40	26.00	34.00	29.00	29.72
10:55	32.00	37.00	36.00	34.99
11:15	39.00	48.00	42.00	43.07
11:25	41.50	50.00	46.00	45.87
11:35	45.00	54.00	51.00	50.01
11:48	47.80	61.80	55.30	55.02
12:00	50.80	72.50	59.00	60.91
12:15	53.80	79.80	64.30	66.13
12:30	56.80	83.70	68.80	69.91
12:56	59.80	90.00	73.30	74.53
13:20	62.80	100.80	77.80	80.71
13:30	81.00	89.00	84.00	84.72
13:56	86.00	93.00	87.00	88.74
14:25	108.00	145.00	140.20	130.97
14:50	119.50	153.00	130.50	134.59

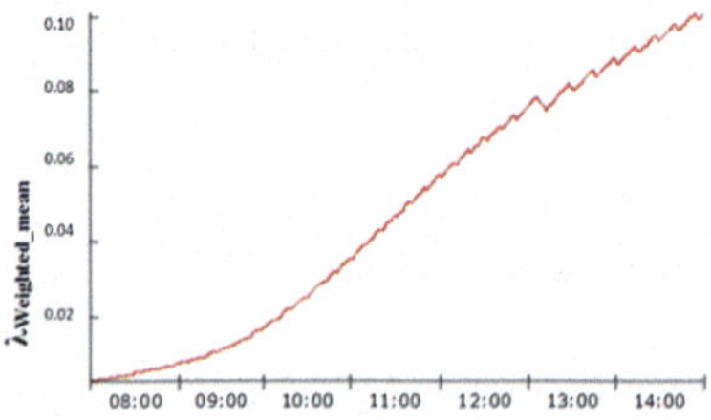

Fig. 9. Failure rate in the work shift time span

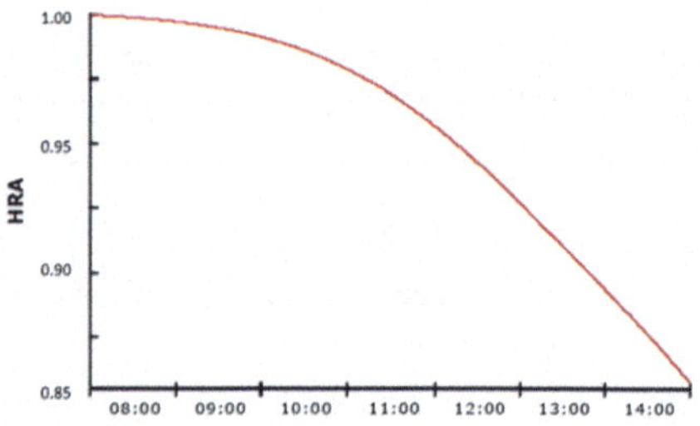

Fig. 10. HRA curve as revealed from probability table

4 Conclusions

In the international scientific literature there is a great lack of knowledge about the influence that the humans can have on the operations management issues. As per the results so far obtained by the researchers, there are several factors that could influence the behavior of human at work. Self reported stress measure were more common then instrumented approach. It is the organizational "climate" the biggest risk factors as it affects the physical, mental and process perspective. Stress is reported as one main cause of error and it requires modelling and context and causality loop. Stress is originated by the mental and visual and physical fatigue the working system acts. So, it is possible to identify a new research focus on behavioral safety based approach. This is focused on main variables that the humans report and those affect the operations at work, i.e., the work-stress. The proposal made apparent the relationships between HF and reliability in system. This is set to understand the perspective of the dynamics actions of events in manufacturing processes.

A comprehensive understanding of how situations present themselves to the operators is an essential input to the HRA analysis. This makes up the context for the personnel's actions; at the same time, condition change by actions. Consequently, managers need to define strategies to adapt system to the human factor is working in. Designers can include, using virtual tools, production and production specification while simulating robotic and tools assistance.

Strategy means choice the right operator at the right task in the correct condition and under changing situations those may be individually and environmentally (internal and external) altered. This paper deals with the presentation of a conceptual framework to identify all the possible factors that influence in different way the work-stress in the field of operations management and in particular around production work environment and related safety facets. After the presentation of the framework to achieve a measure of the humans reliability it is presented in detail each step of the procedure and finally the method it is applied to a real case of study in the automotive environment. The paper demonstrates the effectiveness of the methodology to calculate a curve for the reliability of the humans involved in a generic work environment. This found a mathematic formula for the failure rates of the humans for different tasks. Limitation are related to the method used to build relations. However, the map has the potential to facilitate human factor predictability under imposed hypotheses. Future steps of the actual research work will be the application of the method to other industrial context in the context of co-robotics using sensors for measurables constants and automatic procedure, includes inhuman devices (thinking at i-watches) to capture the dynamics after main variables isolation, and virtual environment to train. In this way the proposal will impact in the designing and operations phase while testing un-commons situations, boosting safety and acting on relevance perceived factors. Co-robotic behavior will change the perception of human work and environment while reporting different weights on the same relevant map. Physical and Mental stress will change consequently with the work process including different main relevancies of the network map as shaping factor for human reliability assessment. Macroergonomics implications for embedding HF aspects will make use of interfaces tested across virtual scenarios.

Appendix

We used questionnaires to collect information from the field. Workers, from the assembly division, were invited to select the appropriate response according to their own sensation at work. Managers were asked for choosing the importance of a particular factor in order to test the variable under investigation. Physician selected the appropriate role among concepts. The answers were elaborated according to the ANP approach with consistency verification. Here is reported a sample of how worker replied. The module is attached in order to guarantee repeatability in the interested analysis. Numerical points are included for illustration - those can be chosen from a list of guided responses.

Check list To EVALUATE
WORK RELATED STRESS

Company	
City (district)	
Activty	
Role/Unit	
Workers in the firm	
Date	

Workers are invited to complete such a checklist. Grouped opinion between wrkers of the same Technological unit can be gotten.
We will proced with simple questions. You can respond to queries included You opinion in a range of values. This could be the resolt of your perception in the context.
This test in ANONYMOUS

You are invited to set a point in a scale from 0 to 9importanza:
(0: not important 9: greatly important

		WEIGHT (please select a poin from the list)
1	HUMAN	8
2	TEAM	2
3	WORK	7
4	ENVIRONMENT	2

HUMAN

	Queries	(select one from suggested options)		How do You retain is the influece of this factor in Your Work context ? 0: not important 9: greatly important	(please select a point from the lista)
1	Sex	F	1	Sex	9
2	Age	40-60	2	Age	9
3	Nationality	Italian	3	Nationality	0
4	Height [m]	1,6 - 1,7	4	Height	7
5	Weight [kg]	80 - 90	5	Weight	7
6	Disability grade	any	6	Disability grade	2
7	Earned Income [€/year]	20.000 - 50.000	7	Earned Income	7
8	Civil Status	Single	8	Civil Status	0
9	No children	0	9	No Children	2
10	Education	Master	10	Education	6
11	Distance Home-Work [Km]	< 1	11	Distance Home-Work	6
12	Means of transport Home - Work	Pullman	12	Means of Transport Home-Work	4
13	Residence	SA	13	Residence	3
14	Experience in the current work [age]	> 30	14	Experience in the current work	6
15	Fatigue level in the tasks You daily perform	Great	15	Fatigue level	9
16	Personal Dissatisfaction	medium	16	Personal Dissatisfaction	7
17	Motivation	Low	17	Motivation	5
18	Do you find particular gap in in competences in what you daily perform	No	18	Competences	6

TEAM

		(select one from suggested options)		How do you retain is the influece of this factor in Your Work context ? 0: not important 9: greatly important	(please select a point from the list)
1	Numerousiness in Team	8 - 15	1	Numerousness in Team	5
2	Work-Team Age	30-40	2	Work Team Age	5
3	Do the workers known table of organization ?	Yes	3	Knowledge of table of organization	4
4	Are You informed of the industrial objectives ?	Yes	4	Knowledge of industrial strategical objecitves	4
5	Are You involved in the safety issues ?	Yes	5	Involvement of the safety plan	4
6	Are there limitation in disciplinary measures ?	No	6	Presence of Disciplinary measures	4
7	Is there a formative plan for workers ?	Yes	7	Knowledge of career opportunities	7
8	Is there moment of strategic discussion ?	Yes	8	Presence of brainstorming, audit	4
9	Is there ethical/behavioural code ?	Yes	9	Presence of ethical code	3
10	Do the workers recevie support if case of work discomfort ?	Yes	10	Support in the case of requirement	8
11	Is clear the definition of role inside the firm ?	Yes	11	Clear definition of roles	8
12	Is there definition of standards for carrier progression ?	Yes	12	Knowledge of standards in carreer	7
13	Is there any premiality ?	Yes	13	Rewards or incentive	9
14	Do You retian the earnings are adequate with Your task ?	No	14	Earnings	9
15	Is there the assence of any crisis and uncertainty in the industrial state ?	No	15	Uncertainty in the firm future	9
16	Is Your work based of work of someone else ?	Yes	16	Indipendence in work	7
17	Are You free to decide how to organize Your work?	No	17	Possibility to freely organize Your work	8
18	Is there clear statement in relation to team behavioral role/organization ?	Yes	18	Knowldge of strategies for team involvement	5
19	Does Your Responsable/Boss pay attention to what You suggest ?	No	19	Partecipation at the strategical decision	7
20	Can the work speak/discute with boss ?	Yes	20	Communication with senior and between workers	8
21	Are you helped if someone in Your organization goes beyond limits of his role ?	No	21	Support in the case of internal conflicts	6
22	Is the absence or minimal perception of querrel and internal conflicts ?	Yes	22	Presence of querrels	4
23	Do you retain Your boss apprecciate (are You in steem) Your work ?	Yes	23	Estimation from Your senior/boss	7
24	Is there absence of competition between people in team ?	Yes	24	Competition between collegues	3
25	Is there absence of physical quarrel ?	No	25	Presence of physical querrels?	5
26	Are You in conflict with Your senior ?	No	26	Conflict with Your senior	4
27	May You freely decide in your task ?	No	27	Autonomous in decision	4

WORK	(select one from sggested options)	How do You retain is the influece of this factor in Your Work context ? 0: not important 9: greatly important	(please select a point from the list)
1 Work Time [hours]	< 8	1 Diration in work time	5
2 Daily/Evening/Nightly shift	Nightly	2 Type of shift (Daily / Nightly)	6
3 No breaks during Your job	1	3 Presence of Breaks	9
4 Cumulative time in your breaks [hours]	< 1	4 Conceived breaks during jobs	6
5 Is there the pssibility to have a meal breaks - Is there a canteen ?	Yes	5 Presence of canteen - meeting point	3
6 Is there the possiblity, if You deman, of flexibilty in task ?	Yes	6 Possibility of flexibility in work time	5
7 Is there the possibility, if You demand, to have part-time work ?	Yes	7 Possibility of part time engagement (is reuired)	3
8 Is there a corporate means of trasport from home to work ?	Yes	8 Possibility to use corporate means of transport	4
9 Is the work good organized ?	Yes	9 Work Organization	4
10 Are the tools of work adquate with Your task ?	Yes	10 Adequate tools	5
11 Are You involved in repetitive tasks ?	Yes	11 Suitable formation	5
12 Is there flexibility in work organization ?	No	12 Flexibility in work time	6
13 Are the human resources adequate with tasks?	No	13 Adequate human engagement	8
14 Is the work amount constant in time ?	No	14 No exceed in work amout	7
15 Is there balancing in work ?	No	15 Balancing in tasks	8
16 Is Your work with low rate repetitiveness ?	No	16 Repetitiveness in task	9
17 Is Your work rhythm fixed in shift ?	No	17 Fixed work rhythm	6
18 Do You decide about Your task ?	Yes	18 Requirement in rapid decision	6
19 Do You retain to use Safety tools ? (Is there safety in system ?)	Yes	19 Satefy in tools/machine	9
20 Is there strictly observation of planned avaiability ?	No	20 Respect of planned time	5
21 Is the planification of hour time clearly defined ?	Yes	21 Defened scheduling	2
22 May You freely decide in breaks during Your work ?	No	22 Possibility to freely manage breaks	9
23 Is there manual material handling during Your job ?	Yes	23 Manual Material handling	1
24 Is the fatigue minimal or absent ?	No	24 Phycical Fatigue	9
25 Is the mental fatigue minimal or absent ?	No	25 Mental Fatigue	8
26 Do You work in team ?	Yes	26 Work in Team	8
27 Are the safety signs clear and easly to detect ?	Yes	27 Safety rule/signs	7
28 Is there adequate maintenance in machine and sercives ?	No	28 Good maintenante plans	9
29 Do You use Personal Protective Equipment ?	Yes	29 Adequate Personal Protective Equipment	9
30 Do You have a suitable work space ?	No	30 Adequate work space	8
31 Are You able to use the correct posture in task ?	No	31 Comfort in task	9
32 Is Your task of great relevance for the industrial item ?	No	32 Importance of position in team	9

ENVIRONMENT Do You retain to be in contact with :	(select one from sggested options)	How do You retain is the influece of this factor in Your Work context ? 0: not important 9: greatly important	(please select a point from the list)
1 Extreme Noice	Yes	1 Noice	9
2 Chemical/cancerogenic Risk	No	2 Chemical/cancerogenic Risk	9
3 Discomfort in Temperature	Yes	3 Discomfort in Temperature	9
4 Not adequate Ventilation	No	4 Ventilation	8
5 Extreme Humidity	No	5 Humidity	9
6 Not adequate Lighting	Yes	6 Lighting	7
7 Vibration	Yes	7 Vibration	5
8 Radiations	No	8 Radiations	9
9 Biological Risk	No	9 Biological Risk	9
10 Presece of powder	No	10 Presece of powder	6
11 Presence of smoky/gas/fumes	No	11 Presence of smoky/gas/fumes	9
12 Risk in fire/explosion	No	12 Risk in fire/explosion	9
13 Not adequate hygienic conditions	No	13 Hygienic conditions	6
14 Not adequate changing/dressing room	No	14 Presence of changing/dressing room	5
15 Sanitary service not clean or not maintained	No	15 Sanitary service	4
16 Atmosferic pressure lower than Your work pressure	No	16 Pressure	8

Main List of Acronyms - Distinguished in categories and, inside those, arranged in alphabetic order

Fundamentals of main methodology
Tools for Human Reliability Assessment
Variables for main functions

FUNDAMENTALs	**ANP**	Analytic Network Process
	CLD	Causal Loop Diagram
	DoE	Design of Experiment
	HF	Human Factor
	HFM	Human Factor Management
	HRA	Human Reliability Assessment
	MALBP	Mixed model Approach for Line Balancing Problem
	OM	Operations Management
	PSF	Principal Shaping Factor
	SD	System Dynamics

(*continued*)

(*continued*)

MAIN VARIABLEs	**a**	allowance rate
	C	Comparison Matrix
	D	Decision vector
	j_{kil}	judgment of decision K compared with I as per expert L
	$\mathbf{L_{max}}$	Fatigue max
	MET	Maximum Endurance Time
	MVC	Maximum Load Contraction
	N	Negative influence in CLD
	P	Positive influence in CLD
	P/N	Positive AND Negative influence in CLD
	$\mathbf{p_i}$	priority of decision factor i
	R	Reliability %
	RA	Rest Allowance
	$\boldsymbol{\lambda}$	failure rate
TOOLs	**ACIH**	Analysis of Consequences of Human Unreliability (translated from French acronyms)
	ATHENA	A Technique for Human Error aNAlysis
	CoCoM	Contextual Control Model
	CREAM	Cognitive Reliability and Error Analysis Method
	DUA	Decision on Unrequired Action
	HEART	Human Error Assessment and Reduction Technique
	HEP	Human Error Probability
	JHEDI	Justified Human Error Data Information
	NARA	Nuclear Action Reliability Assessment
	OOP	Operations Optimization Problem
	OSHA	Occupational Safety and Health Administration
	PRA	Probabilistic Risk Analysis
	PSA	Probabilistic Safety Assessment
	RULA	Rapid Upper Limb Analysis
	RWL	Recommended Weight Limit
	S&C	Snook and Ciriello (S&C)
	SCEBRA	SCEnario Based Risk Assessment
	SIA	Structured Information Analysis
	SLIM-Mud	Success Likelihood Index Methodology, Multi-attribute utility decomposition

(*continued*)

(*continued*)

	SOR	Stimulus, Organism, Response
	SPAR-H	Standardized Plant Analysis Risk–Human reliability analysis
	THERP	Technique for Human Error Rate Prediction

References

Adams, J.D.: Understanding and Managing Stress: A Book of Readings. Universities Associates, San Diego (1980)

Apolloni, S., Fera, M., Macchiaroli, R.: Proposal of a model for chemical risk. SAFE. WIT Transactions, Malta (2007)

Aschoff, J.: Circadian rhythms in man. Science **148**, 1427–1432 (1965)

Basra, G., Kirwan, B.: Collection of offshore human error probability data. Reliab. Eng. Syst. Saf. **61**, 77–93 (1998)

Blackman, H.S., Gertman, D.I., Boring, R.L.: Human error quantification using performance shaping factors in the SPAR-H method. In: 52nd Annual Meeting of the Human Factors and Ergonomics Society (2008)

Borg, G.: Psychophysical scaling with applications in physical work and the perception of exertion. Scand. J. Work Environ. Health **16**, 55–58 (1990)

Brown, M., Tucker, P., Rapport, F., Hutchings, H., Dahlgren, A., Davies, G., Ebden, P.: The impact of shift patterns on junior doctors' perceptions of fatigue, training, worklife balance and role of social. Qual. Saf. Health Care **19**(6), 1–4 (2010)

Cacciabue, P.C.: Human factors impact on risk analysis of complex system. J. Hazard. Mater. **71**, 101–116 (2000)

Colombini, D., Occhipinti, E.: Results of risk and impairment assessment in groups of workers exposed to repetitive strain and movement of upper limbs in various sectors of industry. Med. Lavoro **95**, 233–246 (2004)

Covey, S.R.: The Seven Habits of Highly Effective People. Franklin Covey Co. (1989)

Czeisler, C.A., Gooley, J.J.: Sleep and circadian rhythms in humans. Cold Spring Harb. Symp. Quant. Biol. **72**, 579–597 (2007)

Dawson, D., Reid, K.: Fatigue, alcohol and performance impairment. Nature **338**, 235 (1997)

Deming, W.E.: Out of the Crisis. Massachusetts Institute of Technology, Center for Advanced Engineering Study, Cambridge (1986)

Di Giulio, A., Trucco, P., Invernizzi, F., Citterio, S.: Affidabilità cognitiva dell.operatore umano e sicurezza d.impianto: prospettive di integrazione nei metodi d'analisi. ANIMP (2000)

Di Pasquale, V., Miranda, S., Iannone, R., Riemma, S.: A simulator for human error probability analysis (SHERPA). Reliab. Eng. Syst. Saf. **139**, 17–32 (2015)

Di Pasquale, V., Fruggiero, F., Iannone, R., Miranda, S.: A model for break scheduling assessment in manufacturing systems. Comput. Ind. Eng. **111**, 563–580 (2017)

Dougherty, E.: Human errors of commission revisited: an evaluation of ATHEANA approach. Reliab. Eng. Syst. Saf. **60**, 71–82 (1998)

Dougherty, E.: Is human failure a stochastic process? Reliab. Eng. Syst. Saf. **58**, 209–215 (1997)

El Ahrache, K., Imbeau, D., Farbos, B.: Percentile values for determining maximum endurance times for static muscular work. Int. J. Ind. Ergon. **36**(2), 99–108 (2006)

Fera, M., Macchiaroli, R.: Appraisal of a new risk assessment model for SME. Saf. Sci. **48**, 1361–1368 (2010)

Fera, M., Macchiaroli, R., Fruggiero, F., Lambiase, L.: Risks prioritization in decision making for wind energy investments using analytic network process (ANP). Appl. Eng. Res. **12**, 2567–2574 (2017)

Fruggiero, F., Fera, M., Iannone, R., Lambiase, A.: Work control in balanced DRC systems supported by negotiation procedures between autonomous agents. IFAC-Papers Online **48**(3), 733–740 (2015)

Fruggiero, F., Riemma, S., Ouazene, Y., Macchiaroli, R., Guglielmi, V.: Incorporating the human factor within the manufacturing dynamics. IFAC PapersOnline **49**(12), 1691–1696 (2016)

Fruggiero, F., Fera, M., Lambiase, A., Miranda, S.: What humans act in robotic surgery. In: IEEE International Conference on Industrial Engineering and Engineering Management (2019)

Furuhama, Y., Furuta, K., Kondo, S.: Identification of causes of human errors in support of the development of computer-assisted instruction systems for the plant operator training. Reliab. Eng. Syst. Saf. **47**, 75–84 (1995)

Gallagher, S., Heberger, J.R.: Examining the interaction of force and repetition on musculoskeletal disorder risk: a systematic literature review. Hum. Factors **55**(1), 108–124 (2013)

Gertman, D.I., Blackman, H.S.: Human Reliability & Safety Analysis Data Handbook. Wiley, New York (1994)

Guyton, A.C.: Human Physiology and Mechanisms of Disease, 5th edn. W.B. Saunders, Philadelphia (1992)

Hackman, J.R., Vidmar, N.: Effects of size and task type on group performance and member reactions. In: Marlowe, L. (ed.) Basic Topics in Social Psuchology. Holbrook Press, Boston (1972)

Hallbert, B.P., Gertmann, D.I.: Using information from operating experience to inform human reliability analysis. In: International Conference on Probabilistic Safety Assessment and Management (2004)

Hirschberg, S.: Dependencies, human interactions and uncertainties. Final report of NKS/RAS-470 project, ABB Atom (1990)

Hollnagel, E.: Reliability analysis and operator modelling. Reliab. Eng. Syst. Saf. **52**, 327–337 (1996)

Hollnagel, E.: Cognitive Reliability and Error Analysis Method: CREAM. Elsevier, Amsterdam (1998)

Hollnagel, E.: Looking for errors of omission and commission or The Hunting of the Snark revisited. Reliab. Eng. Syst. Saf. **68**, 135–145 (2000)

Holmberg, J., Hukki, K., Norros, L., Pulkkinen, U., Pyy, P.: An integrated approach to human reliability—decision analytic dynamic reliability model. Reliab. Eng. Syst. Saf. **65**, 239–250 (1999)

Iannone, R., Miranda, S., Riemma, S.: Proposta di un modello per la determinazione automatica delle pause di lavoro in attività manifatturiere a prevalente contenuto manuale. In: Proceedings of ANIMP – Treviso, pp. 46–60 (2004)

Ivanov, P.C., Hu, K., Hilton, M.F., Shea, S.A., Stanley, H.E.: Endogenous circadian rhythm in human motor activity uncoupled from circadian influences on cardiac dynamic. PNAS **104**(52), 20702–20707 (2007)

Johnson, C.: Why human error modeling has failed to help systems development. Interact. Comput. **11**, 517–524 (2014)

Jorgensen, B.W.: Group Size: Its Effects on Group Performance and on Individual Acquisition of Knowledge. Eastern Psychological convention, Washington, D.C. (1973)

Jung, W.D., Yoon, W.C., Kim, J.W.: Structured information analysis for human reliability analysis of emergency tasks in nuclear power plants. Reliab. Eng. Syst. Saf. **71**(1), 21–32 (2001)

Kim, I.S.: Human reliability analysis in the man-machine interface design review. Ann. Nucl. Energy **28**(11), 1069–1081 (2001)

Kirwan, B.: Human error identification in human reliability assessment. Part 1: overview of approaches. Appl. Ergon. **23**(5), 299–318 (1992)

Kirwan, B.: The validation of three human reliability quantification techniques — THERP, HEART and JHEDI: Part 1 — technique descriptions and validation issues. Appl. Ergon. **27**(6), 359–373 (1996)

Kirwan, B.: Human error identification techniques for risk assessment of high risk system. Part 1: review and evaluation of techniques. Appl. Ergon. **29**(3), 157–177 (1998)

Kirwan, B., Gibson, H., Kennedy, R., Edmunds, J., Cooksley, G., Umbers, I.: Nuclear action reliability assessment (NARA): a data-based HRA tool. In: Spitzer, C., Schmocker, U., Dang, V.N. (eds.) Probabilistic Safety Assessment and Management, pp. 1206–1211. Springer, London (2004)

Knisley, J.: Improve employee productivity with custom lighting. Electr. Constr. Maintenance **104**, 34–40 (2005)

Kolus, A., Wells, R., Neumann, W.P.: Production quality and human factors engineering: a systematic review and theoretical framework. Appl. Ergon. **73**, 55–89 (2018)

Konz, S., Johnson, S.: Work Design-Occupational Ergonomics, 6th edn. Holcomb Hathaway, Scottsdale (2004)

Lawler, E.K., Hedge, A., Pavlovic-Vaselinovic, S.: Cognitive ergonomics, socio-technical systems, and the impact of the healthcare information technologies. Int. J. Ind. Ergon. **41**(4), 336–344 (2011)

Savery, L., Luks, J.: The relationship between empowerment, job satisfaction and reported stress levels: some Australian evidence. Leadersh. Organ. Dev. J. **22**(3), 97–104 (2001)

Lazarus, R.S., Folkman, S.: Stress, Appraisal and Coping. Springer, New York (1984)

Lerman, S.E., Flower, D.J., Gerson, B., Hursh, S.R.: Fatigue risk management in the workplace. JOEM **54**(2), 231–258 (2012)

Lombardi, D.A., Folkard, S., Willetts, J.L., Smith, G.S.: Daily sleep, weekly working hours, and risk of work-related injury: US National Health Interview Survey (2004–2008). Chronobiol. Int. **27**(5), 1013–1030 (2010)

Marsden, P., Hollnagel, E.: Human interaction with technology: the accidental user. Acta Psychologica **91**(3), 345–358 (1996)

Matteson, M.T., Ivancevich, J.M.: Controlling Work Stress: Effective Human Resource and Management Strategies. Jossey-Bass, San Francisco, London (1987)

McAtamney, L., Corlett, E.N.: RULA: a survey method for the investigation of work-related upper limb disorders. Appl. Ergon. **24**(2), 91–99 (1987)

McAtamney, L., Corlett, E.N.: Rapid upper limb assessment (RULA). In: Stanton, N., et al. (eds.) Handbook of Human Factors and Ergonomics Methods, 7–11. CRC Press, Boca Raton (2004)

Monsneron-Dupin, F., Reer, B., Heslinga, G., Strater, O., Gerdes, V., Saliu, G., Ullwer, W.: Human-centered modelling in human reliability analysis: some trends based on case studies. Reliab. Eng. Syst. Saf. **58**(3), 249–274 (1997)

Moray, N.: Identifying mental models of complex human-machine systems. Int. J. Ind. Ergon. **22**, 293–297 (1998)

Nembhard, D.A., Osothsilp, N.: Task complexity effects on between-individual learning/forgetting variability. Int. J. Ind. Ergon. **29**(5), 297–306 (2002)

Neumann, W.P., Dul, J.: Ergonomics contributions to company strategies. Appl. Ergon. **40**(4), 745–752 (2009)

Neumann, W.P., Dul, J.: Human factors: spanning the gap between OM and HRM. Int. J. Oper. Prod. Manage. **30**(9), 923–950 (2010)

Niebel, B.W., Freivalds, A.: Methods Standards and Work Design, 11th edn. McGraw Hill, Boston (2003)

Nordin, M., Andersson, G.: Musculoskeletal Disorders in the Work Place: Principle and Practice, 2nd edn. Mosby-Year Book Inc., New York (1997)

Park, K.S.: Human Reliability Analysis: Analysis, Prediction and Prevention of Human Errors. Elsevier Science Publishers, Amsterdam (1987)

Ramaswami, S.N., Agarwal, S., Bhargava, M.: Work alienation of marketing employees: influence of task, supervisory, and organizational structure factors. J. Acad. Mark. Sci. **21**(3), 179–183 (1993)

Rasmussen, J.: Risk management in a dynamic society: a modelling problem. Saf. Sci. **27**(2–3), 183–213 (1997)

Rau, R.: Psychophysiological assessment of human reliability in a simulated complex system. Biol. Psychol. **42**(3), 287–300 (1996)

Reason, J.: Human Error. Cambridge University Press, Cambridge (1990)

Richei, A., Hauptmanns, U., Unger, H.: The human error rate assessment and optimizing system HEROS—a new procedure for evaluating and optimizing the man-machine interface in PSA. Reliab. Eng. Syst. Saf. **72**(2), 153–164 (2001)

Riggs, M.L., Knight, P.A.: The impact of perceived group success-failure on motivational and beliefs and attitudes: a causal model. J. Appl. Psychol. **79**(5), 755–766 (1994)

Rohmert, W.: Problems in determining rest allowances, Part 2: determining rest allowances in different tasks. Appl. Ergon. **4**(3), 158–162 (1973)

Saaty, T.L., Hu, G.: Ranking by eigenvector versus other methods in the analytic hierarchy process. Appl. Math. Lett. **11**(4), 121–125 (1998)

Saaty, T.L.: Fundamentals of Decision Making and Priority Theory with the Analytic Hierarchy Process. RWS Publications, Pittsburgh (2000)

Saaty, T.L.: Decision making with the analytic hierarchy proces. Int. J. Serv. Sci. **1**(1), 83–97 (2008)

Sammarco, M., Fruggiero, F., Neumann, W.P., Lambiase, A.: Agent-based modeling of movement rules in DRC systems for volume flexibility: human factors and technical performance. Int. J. Prod. Res. **52**(3), 633–650 (2014)

Scherkenbach, W.: The Deming Route to Quality and Productivity: Road Maps and Roadblocks. CEEP Press, George Washington University, Washington DC (1986)

Shorrock, S.T., Kirwan, B.: Development and application of a human error identification tool for air traffic control. Appl. Ergon. **33**, 319–336 (2002)

Snodgrass, C.R., Szewczak, E.J.: Human Factors Information System. IGI Global, Hershey (2002)

Snook, S., Ciriello, V.: The design of manual handling tasks: revised tables of maximum acceptable weights and forces. Ergonomics **34**(9), 1197–1213 (1991)

Stanton, N.A., Hedge, A., Brookhuis, K., Salas, E., Hendrick, H.: Handbook of Human Factors and Ergonomics Methods. CRC Press, London (2004)

Steiner, I.D.: Group Process and Productivity. Academic Press, New York (1972)

Sterman, J.D.: Business Dynamics - Systems Thinking and Modeling for a Complex World. McGraw-Hill Higher Education, New York (2000)

Strater, O., Bubb, H.: Assessment of human reliability based on evaluation of plant experience: requirements and implementation. Reliab. Eng. Syst. Saf. **63**, 199–219 (1999)

Sträter, O., Dang, V., Kaufer, B., Daniels, A.: On the way to assess errors of commission. Reliab. Eng. Syst. Saf. **83**(2), 129–138 (2004)

Van Dieen, J.H., Oude Vrielink, H.H.E.: The use of relation between relative force and endurance time. Ergonomics **37**(2), 231–243 (1994)

Vanderhaegen, F.: A non-probabilistic prospective and retrospective human reliability analysis method - application to railway system. Reliab. Eng. Syst. Saf. **71**(1), 1–13 (2001)

Village, J., Greig, M., Salustri, F.A., Neumann, W.P.: Linking human factors to corporate strategy with cognitive mapping techniques. Int. J. Ind. Ergon. **43**(4), 304–313 (2013)

Waters, T., Putz-Anderson, V., Garg, A., Fine, L.: Revised NIOSH equation for the design and evaluation of manual lifting tasks. Ergonomics **36**(7), 749–776 (1993)

Wegner, D.M.: Stress and mental control. In: Reason, S., Fisher, J. (eds.) Handbook of Life, Stress, Cognition and Health, pp. 685–699. Wiley, Hoboken (1988)

Williams, J.C.: A data-based method for assessing and reducing human error to improve operational performance. In: IEEE Conference on Human Factors in Power Plants, Monterey, California, 5–9 June, pp. 436–450 (1988)

Wilson, J.R., Corlett, E.N.: Evaluation of Human Work. Taylor & Francis, London (2005)

Wittingham, R.B.: The Blame Machine: Why Human Error Causes Accidents. Elsevier Butterworth Heinemann, Burlington (2004)

Wright, T.J.: Factors affecting the cost of airplanes. J. Aeronaut. Sci. **3**(4), 22–128 (1936)

Xiang, F., Bingquan, Z., Shengyao, J.: Cognitive model research of nuclear power plant operators. Nucl. Eng. Des. **215**(3), 251–256 (2002)

Yeow, J., Ng, P.K., Sin Tan, K., Chin, T., Yin Lim, W.: Effects of stress, repetition, fatigue and work environment on human error in manufacturing industries. J. Appl. Sci. **14**, 3464–3471 (2014)

Computer-Aided Assessment of Safety Countermeasures for Industrial Human-Robot Collaborative Applications

Fabio Pini(✉) and Francesco Leali

"Enzo Ferrari" Department of Engineering, University of Modena and Reggio Emilia, via P. Vivarelli, 10/1, 41125 Modena, Italy
{fabio.pini,francesco.leali}@unimore.it

Abstract. Human and robot collaboration represents an interesting development direction for traditional industrial robotic solutions. Especially for those cases where the operations are difficult to automate or burdensome for manual execution, the mutual exchange of human sensitivity and robot repeatability represents an effective approach. Nevertheless, industrial robots are poorly involved for collaborative tasks since specific safety countermeasures are required to avoid all the potential hazards. Consequently, the design and assessment of safety solutions represents a fundamental phase to estimate feasibility of industrial collaborative solutions. The presented work proposes a computer-aided approach to identify, assess and optimize the safety systems that enables the collaborative usage of industrial robots. It exploits the capabilities of virtual controller-based offline programming packages to design in advance the safety countermeasures. An initial consistency test validates the response of the selected tool with respect to the safety functionalities. Subsequently, a virtual replica of a potential industrial collaborative solution has been developed. As a result, it has been possible to mimic the behaviour of such a system with respect to Speed and Separation Monitoring collaborative method.

Keywords: Human-robot collaboration · Indutrial robots · Speed and separation monitoring · Computer-aided assessment · Safey countermeasures

1 Safety Assessment for Human-Robot Collaboration in Industrial Applications

Industrial robots are widely recognized as flexible and robust machines with high performances in terms of repeatability and speed. Consequently, their use continuously grows to effectively implement reconfigurable manufacturing systems. Nevertheless, it is still challenging to profitably automate complex tasks, such as those that require high responsiveness and sensitivity with respect to changeable conditions. Human-machine teaming represents a prospective solution [1].

F. Ferraguti et al. (Eds.): HFR 2019, SPAR 12, pp. 186–198, 2020.
https://doi.org/10.1007/978-3-030-42026-0_13

It is based on the synergistic effect of mutual collaboration between human and robot to easily improve the performances of those operations that would be difficult and expensive if performed only by one of the two [2,3]. As evidence of the trend described, in the last years all the main robot manufacturers focused on the development and marketing of both collaborative robots, cobots, and related solutions. Despite the positive trend on cobots [4], use cases in manufacturing systems seem to be still rare [5]. The main motivations are related to the limited speed and payload of cobots, which result in lowered throughput production rate if compared to capabilities of industrial robots.

To extend human-robot collaboration toward industrial manufacturing systems, an interesting solution is to apply safety countermeasures which enable a robust human recognition within the workspace of involved industrial robots. As a result, it is possible to increase the performances of the manufacturing system since the robots will stop or reduce the speed only if the human occupies the shared zone. Conversely, the robots operate at full rate in the other cases. Consequently, the shared zones intended for human tasks could be barrier-free and accessed in parallel to the tasks performed by the robots.

Current available solutions to design safety collaborative strategies between industrial robot and humans are based on dedicated safety controller and sensors, which are connected directly to the industrial robot controller. For instance, industrial robotics manufacturers ABB and FANUC respectively propose the optional safety systems SafeMove [6,7] and Dual Check Safety [8,9], which act as further controlling units above the standard architecture of robot controllers. These units are directly connected to external safety sensors that monitor the robot working area. Commonly used safety sensors are Safety Laser Scanners and Camera Systems; moreover, some solutions exploit proximity sensors applied directly on the robotic arm to detect nearby persons [10], as well as manual guidance equipment to engage a manual guidance scenario [11]. As a result, the main operating methods that involve industrial robot are based on Safety-rated Monitored Stop (SMS), Hand Guiding (HG), and Speed and Separation Monitoring (SSM) [12].

Nevertheless, the selection of specific safety sensors and the implementation of associated safety logics with respect to the process-based robot movements have a direct impact on the efficiency of the human-robot collaborative industrial solution. Consequently, the design of safety countermeasures requires an appropriate assessment of hardware items and software logics. The available literature underlines the importance of assessing the safety of the collaborative application along the design phases; consequently, accordingly to a previous work by the same authors [13], the use of computer aided tools could be extended to evaluate safety countermeasures. The general purpose computer aided design platform DELMIA V5 has been used in [14]; conversely, the dedicated offline programming tool ABB RobotStudio has been proposed by [15]. In the first case it is possible to simulate the system movement with respect to estimated timing for each operation. In the latter case, it is possible to fully replicate the real behaviour of the system since both the robot movements and safety logics are driven by a virtual replica of the robot controller. Consequently, virtual controller-based offline programming packages represent optimal tools to perform pre-empt evaluation tests of robot performances with respect to heterogeneous applications, as demonstrated by different works from many authors [16–21].

With the aim to support the design of effective industrial human-robot collaborative applications, the presented work proposes a computer-aided approach to assess the performance of safety countermeasures for a specific robotic process. The approach relies on the introduction of the virtual controller-based offline programming packages within the design phase to provide an objective validation of prospective collaborative solutions. Specifically, this contribute relies on the offline programming package RobotStudio. Consequently, the next Sect. 2 investigates the capabilities of the selected tool with respect to replicate a basis safety configuration and response of the robot at the change of motion parameters. Subsequently, Sect. 3 describes how to model an industrial solution appropriated for collaborative operations within the offline environment. Two configurations of the solution are evaluated with respect to production capabilities in Sect. 4. Section 5 closes the paper with final remarks and potential directions for future further development.

2 Consistency of Safety Models Within ABB RobotStudio

RobotStudio is the robot offline programming package developed by the ABB company. The capabilities of this tool range from programming of robot paths, configuration of robotic systems, training of robot operators, evaluation and selection of the optimal layout for a specific process, and advanced behavioural simulation to validate robot codes. With respect to safety solutions, RobotStudio provides a specific tool to configure and deploy the logic to manage the response of the SafeMove controller. Figure 1 depicts the graphical interface to set up the SafeMove.

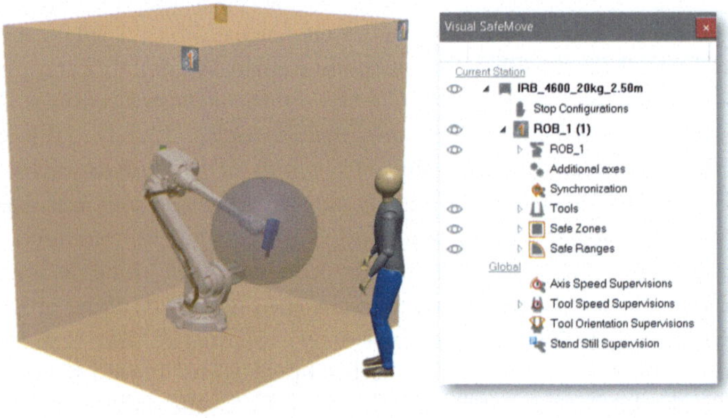

Fig. 1. Graphical configuration of SafeMove logics by means of RobotStudio.

It is possible to reduce the robot workspace by defining a limiting virtual cage and reduced strokes for the selected joints. Moreover, the supervision of stand-still and speeds over robot movements can be deactivated/activated with respect to external digital inputs. Finally, the user can also define the stopping category in case of the infringement of safety conditions. Stop categories 0 and 1 are available. In the first case the robot stops suddenly by immediate removal of power to the actuators using mechanical brakes. Conversely, the second case is a controlled stop that ends the robot motion with electrical power available to the actuators to achieve the stop and then removes of power at the complete stand-still. Based on a virtual replica of the robot controller, RobotStudio allows also to configure the parameters of the controller and the connected logical signal as well. Moreover, it is possible to accurately replicate the movements of the real robot since the virtual controller replicates the same control algorithms that drives the real robot. Consequently, it is possible to define the digital inputs that act directly on the SafeMove functions. As a result, the behaviour of the different robot stopping types and their relation with motion parameters (e.g., End-Effector weight and speed) can be evaluated. Initial tests are performed to evaluate the integrity of virtual data. A simple simulation that stops the robot movements at a specific time by enforcing an emergency safety stop has been defined. The test has been repeated varing the weight of the End-Effector applied on the robot flange and the speed of related tool center point (TCP). The results have been collected using Signal Analyzer, the monitoring tool integrated in RobotStudio. A robot path composed by two joint movements toward two aligned points has been defined. The selected robot is an ABB IRB 4600 with payload of 20 kg and the initial position of the robot has all the joints at 0°, see Fig. 2. Table 1 collects the parameters used for the four initial tests. Specifically, the End-Effector weight assumes the values of 5 and 20 kg, according to the payload of selected robot, and TCP speed varies between 500 and 2000 mm/s. The four tests are the combination of the selected data.

Table 1. Parameters used for initial tests.

Test ID	1	2	3	4
TCP speed (mm/s)	500	2000	500	2000
End-effector weight (kg)	5	5	20	20

To mimic the behaviour of a safety emergency stop, the QuickStop command has been used. Consequently, the final position of the robot along the programmed path is unknown and varies with respect to the instant when the QuickStop actuating (Fig. 2(b)). To replicate the same condition for all the tests, the QuickStop command is timed thanks to logical objects integrated with RobotStudio, defined as SmartComponents. In this case, two Timers have been used. The former pulses the QuickStop digital input after 3 s; the latter activates the StopSimulation function after 4 s. Figure 3(a) depicts the logics developed.

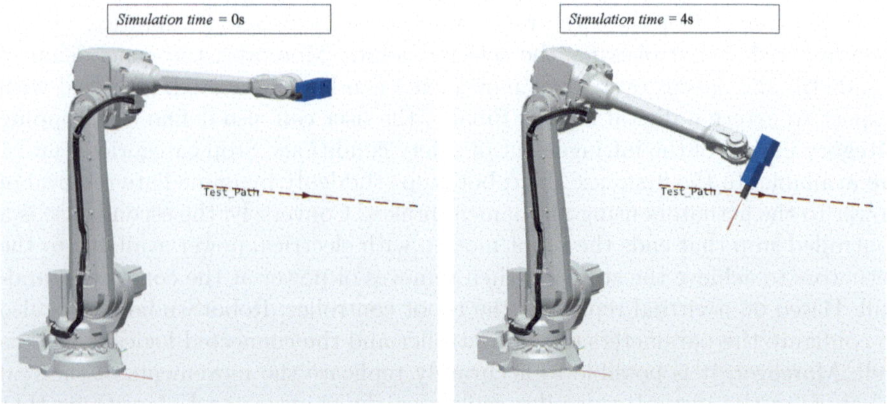

(a) (b)

Fig. 2. Initial and final configurations of the robot used to investigate the consistency of data returned by the simulation. Starting position with all the joints at 0° (a), and final position upon triggering of QuickStop command (b).

Finally, the stopping position of the robot is returned by the data collected with Signal Analyzer, Fig. 3(b); since the selected path mainly involves the rotation of third joint, this initial investigation relies on the position data of this joint.

The results of the consistency tests are collected in Table 2. The pictures are related to the four tests, as mentioned before. The configurations with transparent green refer to the robot position at the instant of QuickStop signal triggering. The coloured configurations represent the final stand-still position of the robot. The two time instants t_{Stop} and t_{Final} identify the two conditions and match those imposed for the Timers, respectively at 3 s and 4 s. The lower rows of the Table 2 provide the angular position of the third joint of the robot, J3, with respect to the two time instants. The data show the capability of simulation tool to calculate the final stopping positions with different motion parameters and confirmation that the dynamic behaviour of the robot is effectively accounted by the virtual controller. Consequently, this first test provides a positive outlook for exploiting the robot offline programming package for subsequent evaluation of the interaction between the human and the robot.

3 Computer-Aided Assessment of Collaborative Safety Countermeasures

The consistency tests proves RobotStudio capabilities in predicting behaviour of the robot with respect to actuation of safety commands. Moreover, the robot offline programming package enables the set-up of advanced safety functions, the design of complex logics and to recreate the communication between different equipment by the exchange of logical signals. As a result, the package can be regarded as a computer-aided tool that could be effectively used to assess

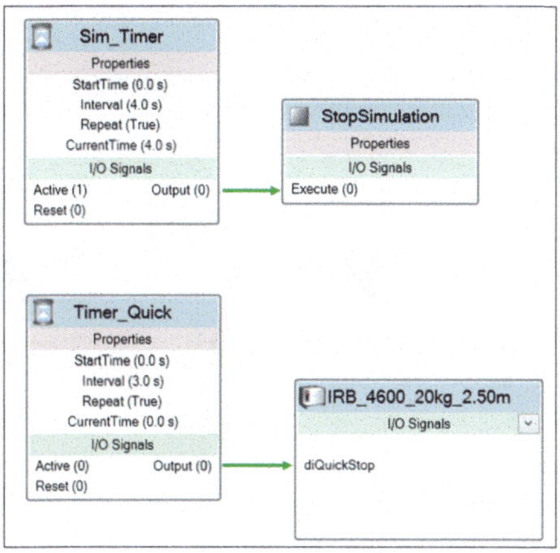

(a)

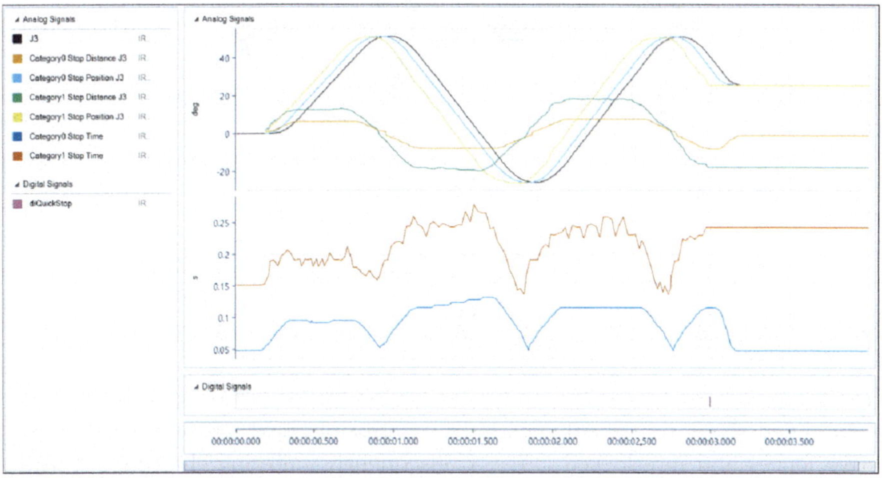

(b)

Fig. 3. Starting model to set up the consistency evaluation. Above, the logic behind the simulation (a); below, a preview of the data recorded by the Signal Analyzer tool (b).

safety scenarios for future collaborative solutions that involve industrial robots and humans. As a proof of the concept, a potential collaborative application is modelled and evaluated within the offline environment. The case study selected is an inspection robotic cell. The robot moves a measurement head over the surface of different components to return the quality data that drive the subsequent operations. Prefatory operations to enable the robotic inspection process, such as loading and positioning of new components, are performed manually by skilled workers. The adoption of a collaborative approach based on SSM method might improve the solution effectiveness. Specifically, the design of the cell can

Table 2. Results of the consistency tests.

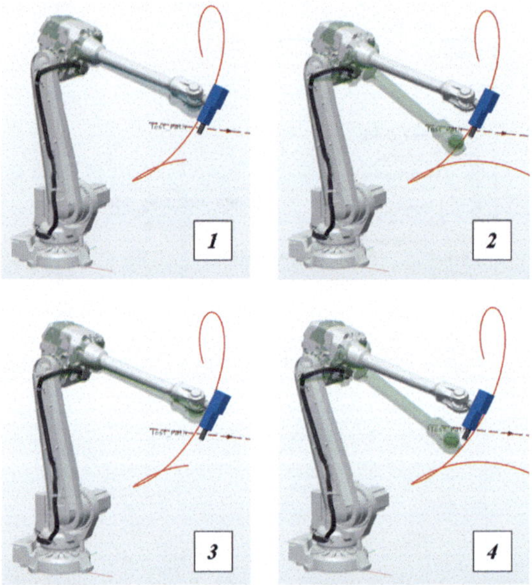

Test ID		1	2	3	4
J3 - t_{Stop}	(deg)	27.31	42.32	26.71	45.67
J3 - t_{Final}	(deg)	24.03	26.09	23.38	31.19

be simplified since it is possible to limit/remove physical barriers, such as protective fences and interlocked gates. Consequently, entering the robot working space does not require to stop the process execution. As a result, it is possible to reduce the robot downtime and simplify the system design.

With respect to development of the SSM method, the robot working area has to be constantly monitored by safety sensors. Commonly used sensors for these purposes are Safety Laser Scanners. It is possible to define different monitoring areas that can be configured as warning or protecting zones. The robot movements can be stopped or slowed down with respect to the distance with the human. The first case is more restrictive since the robot stops and the process is paused until the worker occupies the robot working area. The second case is more flexible and effective since it manages the robot speed in function of the worker location within the robot working area. In both cases, the robot has to reach the expected state before activating respectively the SMS or SSM, otherwise a safety emergency stop is triggered by the safety controller. Consequently, a programmed command has to change the robot speed in advance of activation of the safety checks. The two configuration types of the monitoring zones of laser scanners areas can be used to develop the described logics.

The model of the system has been recreated within RobotStudio environment, as depicted in the Fig. 4. A safety laser scanner has been placed near to the robot base; three control zones have been defined. The large yellow area is a warning zone that imposes a reduction of the robot speed. The little red area near to the robot is a protective zone where the safety controller checks for robot stand-still condition. The intermediate pink area has an hybrid configuration since owns both the protective and warning functionalities. In this area the safety controller checks for robot speed within the select speed limit. In the meantime, it sets to zero to the robot speed in order to reach the stand-still state required in the red area. The component which has to be inspected by the robot is placed inside the red area. Consequently, the process is stopped only in the case the worker moves too close to the robot. To implement the behaviour described, CollisionSensor SmartComponents have been used to recognize the position of the human manikin over the safety areas. These objects recognize the intersection between solid parts (e.g. the models of human manikin and safety areas) and send a digital signal at sensed collision. In the case of the yellow area, the signal activates the LimitSpeed system command that reduces the robot speed to a programmed value. In the case of the pink area, the collision digital input activates also the SSM function of the safety controller. These two consequent check levels allow a controlled reduction of the robot speed and avoid abrupt stop due to quick intervention of SSM. In the case of the red area, the robot stop is performed by dedicated asynchronous robot programmed procedures that interrupts the sequential program flow, known as Interrupts. These are program-defined events executed when the related condition occur, such as the state of a specific signal.

4 Results

The potential collaborative inspection solution has been modelled within the RobotStudio environment. Thanks to right selection of logic blocks and configuration of digital signals it has been possible to replicate the behaviour of the safety system. Figure 5 plots the speeds of robot TCP with respect to simulation time, respectively on the x and y axis. Velocity of internal links of the robot can be included in the safety configuration, but here it has been considered only the TCP speed to simplify the evaluation. Green, yellow and pink colours identified the different collaborative states as follows: worker far from the robot workspace and robot moving at programmed speed, 100 mm/s. Worker approached the robot working space within occupancy of the first warning zone and robot slowed down to reduced speed. Finally, the worker inside the hybrid zone and robot motion inhibited. Despite the TCP speed is not expected to change when the worker is outside the safety area, several fluctuations of the speed can be observed in green area of the plot. This behaviour is not realistic and could be related to data processing by the simulator. The proposed

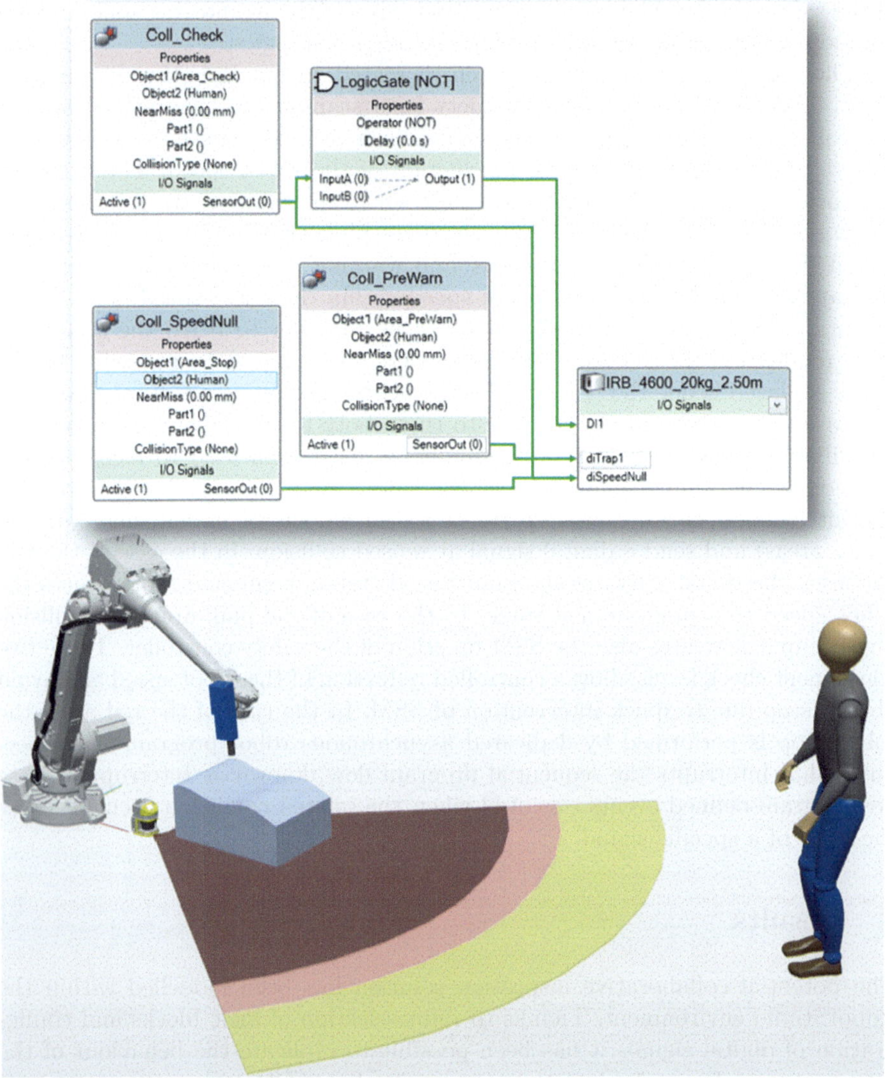

Fig. 4. The model of the collaborative solution for inspection operations. The worker can freely move close to the robot. They have been defined safety areas and related logics to recreate the real behaviour of the system. The block diagram provides the logical architecture behind the simulation based on use of SmartComponents.

model allows further investigations, such as objectively assess the performances of the proposed solution. Figure 6 shows the results related to two different configurations. Configuration (a), on the left side, has been created only for SMS and the robot stops when the human goes inside the safety zone. Conversely, configuration (b) integrates also the SSM, thus it has been improved

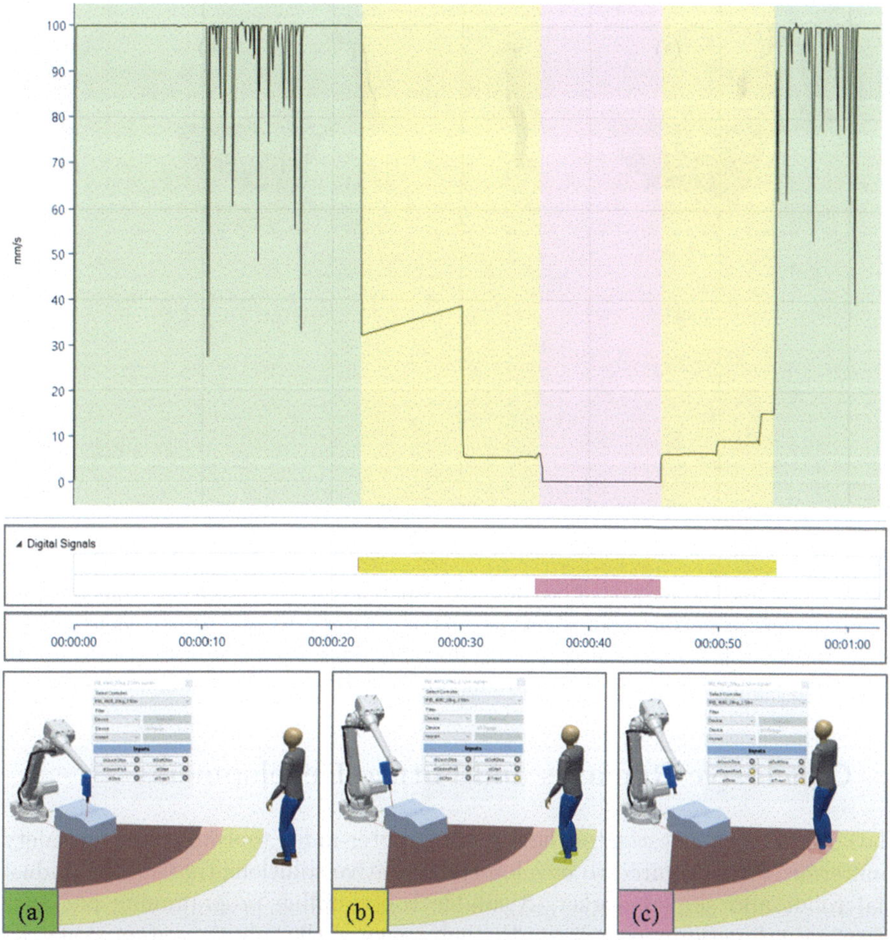

Fig. 5. Results of the simulated collaborative scenarios for the selected use-case.

with one more safety area. The results collected with the Signal Analyzer monitoring tool demonstrate that the collaborative configuration (*b*) reduces the robot stops and that is also possible to quantify the positive gain in terms of production throughput.

Furthermore, the distance between the subsequent zones depends on the relative speed between the human and the robot, and the performances of the selected safety devices. This distance has to be calculated accordingly to ISO/TS 15066:2016 [12]. Consequently, it is possible to impose a specific walking speed for the human and check what is the optimal size for safety zone to perform a controlled speed reduction.

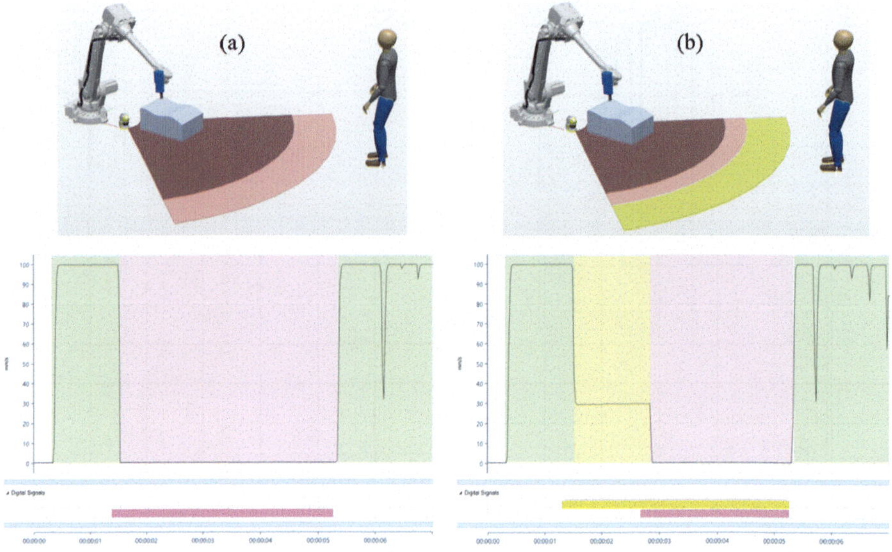

Fig. 6. Objective comparison of two different collaborative configuration. Configuration (a), on the left, stops the robot when the human occupies the safety area; configuration (b), on the right, has an intermediate zone that slow down the robot before the complete stop. The plots demonstrate better production performances for configuration (b).

5 Conclusive Remarks and Future Developments

The current work investigates the use of computer-aided tools to assess the safety countermeasures required to develop collaborative solutions that involve industrial robot and skilled worker. Available robot offline programming packages offers many functions to replicate the behaviour of the robotic system along the operations of a specific process. In particular, the packages based on the virtual replica of the robot controller provide advanced functionalities such as the configuration of safety logics and the design of complex models. As a results, these tools might represent a valid support also to asses the safety solution that effectively implements industrial human-robot collaborative applications and can be used along the design phase of such a systems.

The robot offline programming package selected in this work is ABB RobotStudio. An initial evaluation stresses the consistency of the data returned with respect to simulating of the usual emergency actions, such as the actuation of an emergency stop. The results demonstrate good capability of the tool to replicate the real behaviour of the robot that reaches different stopping positions in function of selected motion parameters. Subsequently, the capability of the tool to replicate a collaborative scenario has been evaluated. The use case selected has been a robotic workcell for inspection operations. The collaborative solution is based on SSM and SMS functions, which have been configured on the SafeMove controller, and integrated in RobotStudio. As a result, it has been possible to define a model of the collaborative solution and to positively test its response with respect to collaborative actions. Consequently, the model can also be used to investigate alternative solutions in order to improve the effectiveness of the final solution and enforce the safety of collaborative tasks, such as including the speed of internal links of the robot in the safety configuration.

Nevertheless, the present work requires further investigations to fully demonstrate the capabilities of computer-aided tools. Proper tests have to be performed on real industrial robots to validate the consistency of the simulated data with respect to the data collected on the real robot.

References

1. Fast-Berglund, Å., Palmkvist, F., Nyqvist, P., Ekered, S., Åkerman, M.: Evaluating cobots for final assembly. Procedia CIRP **44**, 175–180 (2016). https://doi.org/10.1016/j.procir.2016.02.114
2. Bragança, S., Costa, E., Castellucci, I., Arezes, P.: A brief overview of the use of collaborative robots in industry 4.0: human role and safety. Stud. Syst. Decis. Control **202**, 641–650 (2019). https://doi.org/10.1007/978-3-030-14730-3_68
3. Malik, A., Bilberg, A.: Digital twins of human robot collaboration in a production setting. Procedia Manuf. **17**, 278–285 (2018). https://doi.org/10.1016/j.promfg.2018.10.047
4. Bloss, R.: Collaborative robots are rapidly providing major improvements in productivity, safety, programing ease, portability and cost while addressing many new applications. Ind. Robot **43**(5), 463–468 (2016). https://doi.org/10.1108/IR-05-2016-0148
5. Oberc, H., Prinz, C., Glogowski, P., Lemmerz, K., Kuhlenkötter, B.: Human robot interaction - learning how to integrate collaborative robots into manual assembly lines. Procedia Manuf. **31**, 26–31 (2019). https://doi.org/10.1016/j.promfg.2019.03.005
6. Kock, S., Bredahl, J., Eriksson, P., Myhr, M., Behnisch, K.: Taming the robot. ABB Rev. **4**, 11–14 (2006)
7. ABB - SafeMove2 - Protecting operators and enhancing robot safety. https://new.abb.com/products/robotics/controllers/irc5/irc5-options/safemove-2. Accessed 26 July 2019
8. Goldsberry, C.: Worker safety remains an important ingredient in automated systems. Weld. Des. Fabr. **81**(2) (2008)
9. FANUC - Dual Check Safety DCS. https://www.fanuc.eu/uk/en/robots/accessories/safety. Accessed 26 July 2019
10. Byner, C., Matthias, B., Ding, H.: Dynamic speed and separation monitoring for collaborative robot applications - concepts and performance. Robot. Comput. Integr. Manuf. **58**, 239–252 (2019). https://doi.org/10.1016/j.rcim.2018.11.002
11. Michalos, G., Kousi, N., Karagiannis, P., Gkournelos, C., Dimoulas, K., Koukas, S., Mparis, K., Papavasileiou, A., Makris, S.: Seamless human robot collaborative assembly - an automotive case study. Mechatronics **55**, 194–211 (2018). https://doi.org/10.1016/j.mechatronics.2018.08.006
12. ISO/TS 15066:2016 - Robots and robotic devices - Collaborative robots. ISO (2016)
13. Pini, F., Ansaloni, M., Leali, F.: Evaluation of operator relief for an effective design of HRC workcells. In: IEEE International Conference on Emerging Technologies and Factory Automation (ETFA 2016), p. 7733526 (2016). https://doi.org/10.1109/ETFA.2016.7733526

14. Wadekar, P., Gopinath, V., Johansen, K.: Safe layout design and evaluation of a human-robot collaborative application cell through risk assessment - a computer aided approach. Procedia Manuf. **25**, 602–611 (2018). https://doi.org/10.1016/j.promfg.2018.06.095
15. Lehment, N., Boca, R.: Performance impact assessment for SSM using simulated stations. In: 50th International Symposium on Robotics (ISR 2018), pp. 131–134 (2018)
16. Holubek, R., Ružarovský, R., Sobrino, D.: An innovative approach of industrial robot programming using virtual reality for the design of production systems layout. In: Trojanowska, J., Ciszak, O., Machado, J., Pavlenko, I. (eds.) Advances in Manufacturing II. Lecture Notes in Mechanical Engineering, pp. 223–235. Springer, Cham (2019). https://doi.org/10.1007/978-3-030-18715-6_19
17. Obal, P., Burghardt, A., Kurc, K., Szybicki, D., Gierlak, P.: Monitoring the parameters of industrial robots. In: Hanus, R., Mazur, D., Kreischer, C. (eds.) Methods and Techniques of Signal Processing in Physical Measurements. Lecture Notes in Electrical Engineering, pp. 230–238. (2019). https://doi.org/10.1007/978-3-030-11187-8_19
18. Forgó, Z., Hypki, A., Kuhlenkötter, B.: Gesture based robot programming using ROS platform. In: 50th International Symposium on Robotics, ISR 2018, pp. 455–461 (2018)
19. Casanelles, R., Vilacoba, D., Weber, P.: Towards high performance in press automation: an ABB view. In: Proceedings of the IEEE International Conference on Industrial Technology, p. 7125573 (2015). https://doi.org/10.1109/ICIT.2015.7125573
20. Pini, F., Leali, F., Ansaloni, M.: Offline workpiece calibration method for robotic reconfigurable machining platform. In: 19th IEEE International Conference on Emerging Technologies and Factory Automation, ETFA 2014, p. 7005246 (2014). https://doi.org/10.1109/ETFA.2014.7005246
21. Connolly, C.: Technology and applications of ABB RobotStudio. Ind. Robot **36**(6), 540–545 (2009). https://doi.org/10.1108/01439910910994605

Author Index

F. Ferraguti et al. (Eds.): HFR 2019, SPAR 12, pp. 199–200, 2020.
https://doi.org/10.1007/978-3-030-42026-0

The manufacturer's authorised representative in the EU is Springer Nature Customer Service Centre GmbH, Europaplatz 3, 69115 Heidelberg, Germany. If you have any concerns regarding our products, please contact ProductSafety@springernature.com

Printed and bound by CPI Group (UK) Ltd, Croydon, CR0 4YY
30/09/2026
02225449-0004